Bees, Wasps, and Ants

Bees, Wasps, and Ants

The Indispensable Role of Hymenoptera in Gardens

Eric Grissell

Timber Press
Portland • London

Published in 2010 by Timber Press, Inc.

The Haseltine Building
133 S.W. Second Avenue, Suite 450
Portland, Oregon 97204-3527
timberpress.com

2 The Quadrant
135 Salusbury Road
London NW6 6RJ
timberpress.co.uk

Second printing 2012
Printed in China

Library of Congress Cataloging-in-Publication Data
Grissell, Eric.
    Bees, wasps, and ants : the indispensable role of Hymenoptera in gardens / Eric Grissell.
       p. cm.
    Includes bibliographical references and index.
    ISBN 978-0-88192-988-1
    1. HymenopteraPopular works. 2. BeesPopular works. 3. Wasps—Popular works. 4. Ants—Popular works. I. Title.
    QL565.G72 2010
    595.79—dc22

                                                            2009051603

A catalog record for this book is also available from the British Library.

*This work is dedicated to
Ellsworth Hagen, my high school biology
teacher, who was the first to encourage
what became a lifelong study of insects, and to
Richard Bohart, my major professor and mentor,
who was the first to encourage my study of
Hymenoptera, which became my profession.
Both were examples of the effect that
caring and energetic teachers have upon
impressionable, young students.*

Contents

Color photographs begin on pages 97, 181, and 245

Preface

AS WITH MANY FOLKS, my introduction to the subject of bees, wasps, and ants began somewhere near the age of five or six while innocently running barefoot across the lawn at a friend's house. I stepped on a honey bee, herself innocently gathering nectar from a patch of clover. Two innocents, each having fun in our individual ways, and each losing our innocence at the same moment: me, by learning the notion that some insects sting, and the bee, by learning—albeit briefly—that honey bees self-destruct in their own defense (this is unique to honey bees). Obviously my learning experience was more rewarding than the bee's, but it did instill a fear of stinging insects, which nearly 60 years later is still fresh in my mind.

It would seem odd, you might agree, that having gained an early fear of being stung, I became a professional entomologist. Naturally I retained a childhood fear of stinging insects, but never having developed the intelligence to avoid them, and defying common sense, I chose to study one of the orders of insects, the Hymenoptera, that contain the nastiest stinging insects in the world, namely wasps, bees, and ants. As a result, since my earliest childhood I've been stung by about every member of the group commonly found in the United States and actually lived to write about it.

If I were to walk into a crowd of people and ask the first person I met to name five Hymenoptera, chances are I would receive a befuddled stare at best, or a slap in the face at worst. Hymenoptera is the name of an order used by entomologists for what the public refers to as bees, wasps, and ants. There are so many insects and such a jumble of minute technical differences among them that entomologists have developed thousands of terms and nuances to discuss the million known insects. Although some orders of insects are simply referred to by a single common name, such as beetles (Coleoptera), flies (Diptera), or termites (Isoptera), there is no single common name for Hymenoptera. Thus, when communicating with the public it is common to speak of bees, wasps, and ants, but it would be easier in the long run if we could learn to speak of Hymenoptera, or "hyms" as they are sometimes called. It is a single word that represents a huge amount of information in a simple way. Incidentally, those of us who study the group rarely use the term *hymens* when discussing them for reasons of decorum, but the word does plays a role in relation to the derivation of the ordinal name.

In many respects the use of the big three terms, namely *bees, wasps*, and *ants*, has its difficulties. Whereas the notion of *ant* probably means the same to everyone, the terms *bee* and *wasp* are basically muddled in the minds of the general populace and are mostly used indiscriminately. Recently, an episode of the television show *The Simpsons* (entitled "The Burns and the Bees") attempted to exploit the hysteria surrounding the sudden nationwide disappearance of honey bees. Although merely a cartoon, the show managed to get some aspects of reality correct, except they showed honey bees living in a hornet's (a wasp's) nest hanging from a tree—a minor mistake to some, but typical. To an entomologist this is the equivalent of someone talking about butterflies when they really mean dragonflies. Bees and wasps are not the same, but as this is simply an introduction to more detailed aspects yet to come, we needn't dwell on that subject just yet.

When used correctly, the term *bee* is used by most folks simply to refer to the social honey bees and bumble bees. Yet there are thousands of solitary bees in the world, and dozens that go unnoticed in our gardens simply because they are easily overlooked. Bees, although nearly 20,000 species strong worldwide, are not especially diverse in their biological needs (at least compared to wasps), so their role in the environment is relatively easier to characterize than most other Hymenoptera. With few exceptions bees are flower visitors, gathering pollen and nectar for their offspring, although they do have some deviant behaviors perhaps not expected by the gardener.

When it comes to recognizing Hymenoptera, the most baffling and misunderstood group would have to be the wasps. This group comprises tens of thousands (or even hundreds of thousands if you count the species yet likely to be discovered) of insects mostly not known or seen by the general public. The word *wasp* generally connotes only a few irritating species of the social stinging sorts: yellow jackets, hornets, and paper wasps. But by far the majority of wasps are of the parasitic and solitary kinds, which feed on nearly every other insect imaginable in fascinating and horrifying ways (Figure 1). Then, too, there are woodwasps, horntails, and sawflies that feed on and in plants. The word *wasp*, then, is fraught with misconceptions, confusion, and misuses. In the chapters that follow, I have attempted to clarify the meaning of the word and to categorize wasps in a way that reflects not only their relationships to each other but also their biological position in the natural world.

Examples of nearly every kind of bee, ant, and wasp can be found somewhere in a garden, but they work in mysterious and wondrous ways, largely unknown to those of us who tend the land. This book, then, is an exploration of the lives of Hymenoptera, about which we—and I include entomologists as well as myself here—know so little. It is an openly impassioned plea for acceptance of life forms that are vastly greater in importance to us than most humans would imagine and hugely necessary to the world in which we live. And it is a suggestion that we extend an invitation to bees, wasps, and ants to move into and protect our gardens as a way of maintaining a degree of balance in our lives.

Having worked with Hymenoptera my entire adult life, I know the first question to arise in the minds of reasonable people will be: "Why in the world do we want these horrific, painful thugs living in our garden?" I imagine that the words used might be slightly stronger than those, as I myself have used a number of such colorful words during a life spent working among the Hymenoptera. Still, in the grand scheme of things, bees, wasps of all sorts, and ants are among the most important insects on our planet.

Before proceeding, however, I must offer a word of warning and an apology. This is not a field guide to the identification of Hymenoptera. Identification of these creatures, even to the family level, is not for the faint of heart because many appear very much alike even though they are totally unrelated (Figure 2). In addition, numerous specialist terms are needed as well as a fairly high powered dissecting microscope. Even a hand lens won't do when counting the antennal segments, for example, of a wasp that is $1/16$ inch (1.5 mm) in length. The best I can do within the limits of this book is to present illustrations of the various kinds likely (and unlikely) to be seen by gardeners as they flit about their gardens—both the hyms and the gardeners, that is.

Within each chapter I provide a few comments on other sources to visit, both books and websites. Currently the most widely used identification aid among beginning students is *Borror and DeLong's Introduction to the Study of Insects* (Triplehorn and Johnson 2005), but this is no field guide! The tome includes keys for all orders and families of North American insects, including Hymenoptera. Those rising to the ultimate achievement in hymenopteran identification would best consult a tome entitled *Hymenoptera of the World: An Identification Guide to Families* (Goulet and Huber 1993). The work is nearly 700 pages in length with hundreds of black-and-white illustrations. Needless to say it verges on overkill and is not designed as a quick reference. At the opposite extreme are a few illustrated field guides to insects, but none are devoted entirely to Hymenoptera. Unfortunately all such guides are necessarily limited in scope and purpose because of the overwhelming number of insects. My suggestion is that you examine field guides at a local bookstore to see if any might be helpful for your particular purpose or for your particular region.

For North America I hesitate to recommend any one guide in particular, but the *Kaufman Field Guide to Insects of North America* (Eaton and Kaufman 2007) stands out among the rest because it is well illustrated with iconographic images of Hymenoptera. A list of additional potentially useful field guides include: *National Wildlife Federation Field Guide to Insects and Spiders and Related Species of North America* (Evans 2007), *A Field Guide to Insects* (Borror and White 1998), *Peterson First Guide to Insects of North America* (Leahy 1998), *Simon & Schuster's Guide to Insects* (Arnett and Jacques 1981), and *National Audubon Society Field Guide to North American Insects and Spiders* (National Audubon Society 1980). One additional guide to general insects especially for gardeners is *Garden Insects of North America: The Ultimate Guide to Backyard Bugs* (Cranshaw

2004), which is a useful photographic encyclopedia of garden insects, including some Hymenoptera.

European field guides are a bit more difficult to consider as many refer to smaller portions of the hymenopteran fauna (i.e., families or genera) based on country or region. The most helpful general guides are *Collins Field Guide: Insects of Britain and Northern Europe* (Chinery 1993) or *Field Guide to Insects of Great Britain and Northern Europe* (Gibbons 1996). Another useful guide is *Bienen, Wespen, Ameisen: Hautflügler Europas* (Bellmann 2005) in German and its French translation *Guide des abeilles, bourdons, guêpes et fourmis d'Europe* (Bellmann 2009). Gardeners in the United Kingdom are fortunate in having a number of guides, but they not always easy to obtain. Some specific guides are listed at the Bees, Wasps, and Ants Recording Society (BWARS, see list of useful websites). Printed guides include The Naturalists' Handbook Series guides to *Bumblebees* (Prys-Jones and Corbet 1991), *Plant Galls* (Redfern and Askew 1992), *Solitary Wasps* (Yeo and Corbert 1995), and *Ants* (Skinner and Allen 1996). An up-to-date publication on bumble bees was written by Edwards and Jenner (2009). An older series of numerous guides to individual families was published as *Handbooks for the Identification of British Insects*, but these are largely out of print and not especially easy to find. Another older work still in print is *The Hymenopterist's Handbook* (Betts and Laffoley 1986) with all sorts of information and identification aides for the fauna of Britain.

For those of you who are really stung by the Hymenoptera bug (to mix metaphors), there are a few highly intense field/lab courses available that might appeal to your sense of completeness. The three that come to mind are a general course on the order and two courses specifically devoted to ants and bees. The "Parasitic Hymenoptera Short Course" has been given every other year for more than two decades by a consortium of the Systematic Entomology Laboratory (USDA), the Smithsonian Institution, and the University of Maryland. It has recently been expanded and renamed "The Hym Course." Its location changes from year to year, but sometimes it is held at the American Museum of Natural History's Southwestern Research Station in Portal, Arizona. The "Ant Course," run every summer, and the "Bee Course," run every other summer, are both conducted at the Southwestern Research Station as well. Information on these courses is given in the list of useful websites. An annual course entitled "Taxonomy and Biology of Parasitic Hymenoptera" was run by The Natural History Museum (London) and Imperial College (Silwood Park) for several years, but is currently dormant for lack of funds.

A Note About Terminology

- •

THROUGHOUT THIS BOOK I use as few scientific terms as possible simply to make the reading go a bit easier. Technical terms are defined upon their first use and are included in the index as well. There are a few collective terms, however, that are used rather freely, so it would be best to define them up front. Although I use common names where possible, most individual species of bees, wasps, and ants have no common name so they are often referred to simply by some shortened version of their family name. Some families do have a common collective name, for instance, spider wasps for the family Pompilidae.

In speaking about animals, especially insects because there are so many of them, biologists use categories that are precise so that we can communicate a maximum amount of information with a minimum number of words. For example, the terms *order, superfamily, family, genus,* and *species* are used simply to lump huge assemblages of related things together in a nested set of categories. In theory each of these terms refers to a biologically lower category, so that an order is a group of related superfamilies, a superfamily is a group of related families, a family is a group of related genera (singular: genus), and a genus is a group of related species. The term *species* is both singular and plural and is sometimes abbreviated when writing about a species that we can't (or don't) identify. For example, "*Apis* sp." refers to a single undetermined honey bee species of the genus *Apis,* whereas "*Apis* spp." refers to several species. If we wrote out the term *species* in these instances, the reader would not know if we were talking about one bee or many bees.

With species, the problem becomes even more complicated, because they are generally agreed to be groups of individuals that do not successfully interbreed with other related groups of individuals (other species) in nature. In reality, lots of so-called species (both plant and animal) will interbreed given the chance and sometimes they produce viable offspring or hybrid swarms (in plants). There are many terms to cover these categories, including *subspecies* (if they are geographically distinct populations), *forms, varieties,* or *hybrids.* Thankfully we rarely need to sink to this level of detail, but the terms from species to order are used frequently in this book.

Acknowledgments

I WOULD LIKE TO THANK the following individuals for help in preparing this work. Without them I would likely never have finished. If errors are found within its covers, I am totally responsible. In spite of the fact that I've studied Hymenoptera nearly my entire life, it seems I am easily confused and frequently mystified by the vagaries and complexity found within the group. The more I think I know, the less I actually do. So I can only hope that my colleagues and friends will forgive me if I have misinterpreted their input.

Celso Azevedo (Departamento de Biologia, Universidade Federal do Espirito Santo, Maruípe, Brazil), for information on numbers of Bethylidae used in the table of hymenopteran families; Matthew Buffington (research entomologist, Systematic Entomology Laboratory, U.S. Department of Agriculture [USDA], Washington, D.C.), for providing information about gall wasps and their relatives and for images; Robert Carlson (retired, Communications and Taxonomic Services Unit, Systematic Entomology Laboratory, USDA, Fairfax, Virginia), for information about ichneumonids and many images used in the book; James Carpenter (curator of Hymenoptera, American Museum of Natural History, New York), for identification of several predatory wasp photos; Susan E. Ellis (USDA, Animal and Plant Health Inspection Service, Plant Protection and Quarantine, Carneys Point, New Jersey), for several images; Michael Gates (research entomologist, Systematic Entomology Laboratory, USDA, Washington, D.C.), for providing images of tiny wasps and training me in the use of the EntoVision Imaging Suite in his lab; Carll Goodpasture (Gjettum, Norway), for constant encouragement, reviewing parts of the manuscript, suggesting many important improvements, and providing many of the images for this book; Terry Griswold (research entomologist, USDA, Agricultural Research Service, Bee Biology and Systematics Laboratory, Utah State University, Logan), for confirmation of some bee identifications; Ron Hodges (research entomologist, retired, Systematic Entomology Laboratory, USDA, Eugene, Oregon), for help with identification of lepidopteran larvae; Norman Johnson (professor, Department of Entomology, Ohio State University, Columbus), for the use of several wasp images; Lynn Kimsey (professor, Department of Entomology, University of California, Davis), for information on Chrysididae and Tiphiidae used in the table of hymenopteran families; George Melika (entomologist, Pest Diagnostic Laboratory, Plant Protection and Soil Conservation Directorate of County Vas Tanakajd, Hungary), for identification of several of the gall wasp photos; Arnold Menke (research entomologist, retired, Systematic Entomology Laboratory, USDA, Bisbee, Arizona), for identification of several predatory wasps; John S. Noyes (research entomologist, retired, Department of Entomology, The Museum of Natural History, London), for providing information on Chalcidoidea used in the table of hymenopteran families; Juliet Osborne (Deputy Director, Centre for Soils and Ecosystem Function, Rothamsted Research, Harpenden) for suggesting several websites; Frank Parker (research entomologist, retired, USDA, Agricultural Research Service, Bee Biology and Systematics Laboratory, Utah State University), for unpublished information used in chapter 4, the identification of several bee images, and an image of bee nests in his garden; Wojciech Pulawski (curator, Department of Entomology, California Academy of Sciences, San Francisco), for identification of several of the predatory wasp photos; Dave Smith (research entomologist, retired, Systematic Entomology Laboratory, USDA, Falls Church, Virginia), for information about sawflies and reviewing and improving the chapter on his favorite subject; Robbin Thorp (professor emeritus, Department of Entomology, University of California, Davis); David Wagner (associate professor, Department of Ecology and Evolutionary Biology, University of Connecticut, Storrs), for identification of lepidopteran larva; David Wahl (curator, American Entomological Institute, Gainesville, Florida), for information about ichneumonids; Robert Wharton (professor, Department of Entomology, Texas A&M University, College Station), for information about plant-feeding braconids; and Dicky Yu (research analyst, Department of Entomology, University of Kentucky, Lexington), for providing information on Ichneumonoidea used in the table of hymenopteran families.

I especially wish to thank Carll Goodpasture and Robert W. Carlson, who provided many of the images in this book. Collectively a number of images were provided by Insect Images, a joint project of The Bugwood Network; USDA Forest Service; and The University of Georgia, Warnell School of Forestry and Natural Resources and the Department of Entomology in the College of Agricultural and Environmental Sciences. I am thankful to Elizabeth Carlson (Center for Invasive Species and Ecosystem Health, The University of Georgia, Tifton), who was most helpful in securing permission to use these images. Each image provided by these contributors is attributed to its creator. Images without attribution were taken by the author.

Finally, I am especially grateful to Lisa DiDonato Brousseau for her expert editorial skills, which have proven both insightful and humbling.

PART 1

AN OVERVIEW OF BEES, WASPS, AND ANTS

W HEN DISCUSSING HYMENOPTERA, superlatives falter in the face of reality. The smallest insect and the largest brood of any known insect are found in the parasitic wasps. The largest insect egg, the most spectacular matings, and the largest sexual orgies are found in the bees. The greatest host specificity of any insect is found in fig wasps. Among insects, the greatest heat tolerance and most toxic insect venom are found in the ants. Among all animals the fastest anatomical structure ever recorded in the animal kingdom is also found in an ant. There is evidence that a single ant colony in the Mediterranean extends underground more than 3600 miles and that another in Japan houses more than 300 million workers and 1 million queens. Also in Japan, an invasive force of 30 giant hornets can wipe out an entire colony of 30,000 honey bees. Such feats boggle the mind, but when it comes to Hymenoptera almost anything seems possible, including several Pulitzer Prizes, a Nobel Prize, and even the creation of the Kinsey Institute for Research in Sex, Gender, and Reproduction. For the intrepid reader, all will be revealed in its time.

Numerically speaking, Hymenoptera is one of the largest insect orders in the world. Only the beetles (Coleoptera) outnumber them by a large margin. For North America the figure for bees, wasps, and ants is about 22,000 species and for Europe about 15,000 species. This is probably a lot more bees, wasps, and ants than the average gardener cares to imagine! Determining the exact world figures is a bit more challenging because numbers vary from source to source and day to day, but based on the information presented in the table of hymenopteran families, a decent estimate would be somewhere around 150,000 described species. The hymenopterans, however, have never been as popular a group to study as have the beetles or butterflies, so there are untold thousands of species awaiting discovery— anywhere from 300,000 (Gauld and Gaston 1995) to nearly 2.5 million (Stork 1996) according to various authors based on established sampling routines. All these species are distributed in 89 families known throughout the world, with 75 represented in North America.

In 1983 Jennifer Owen published *Garden Life* in which she summarized a decade of insect trapping in her suburban Leicester, England garden. Her totals for Hymenoptera were 46 species of bees, 40 species of predatory wasps, and 553 species of parasitic wasps. She admitted that only selected species of parasites were identified due to the difficulty in doing so. In spite of the enormous numbers of Hymenoptera species in nature and our gardens, about which most of us are unaware, their reputation is based largely on the honey bee, a few hornets and yellow jackets, and the omnipresent ant (Figures 3, 4). To add insult to injury, these few examples of the order form the basis for a reputation that is generally not portrayed with positive press. As children growing up in the 1950s, we watched the horrifying films of the Old West in which innocent victims were buried in mounds of ants to be eaten alive, flesh ripped away sliver by tiny sliver. Whether true or not, similar horror stories of today are related through newspaper and television. Not long ago I read a report in the *El Paso Times* about a chap who was walking to the local grocery to buy a pack of cigarettes just as some playful children knocked a "killer beehive" from a tree. As a result, 15 people were sent to the hospital, and the unlucky chap died from a heart attack induced by innumerable stings to the head. (Could this be additional evidence that smoking is dangerous to one's health?) Tales of similar threats to life and limb, whether by fire ant, killer bee, or angry wasp, though rare, are not unheard of and generally result in local banner headlines of the negative kind (Figure 5).

Working with the general public as I have done over the years, one of the most difficult tasks I'm faced with is defending the offenses of some of these less popular members of the Hymenoptera. Generally the problem revolves around some encounter with a "bee," which is what many people refer to when they really mean wasp. On one occasion, when I was in charge of an insect fair booth on the National Mall, a fellow walked up to me and asked,

"What should I do about bees?"

"What kind of bees?" I rejoined, as it is a well-known entomological fact that one can scarcely do anything about insects until one knows what kind of insect it is.

"Bees that live in the ground," he replied. "Lots of them. Black and yellow."

It's difficult to address these sorts of questions because the information often does not describe a known situation, or at least one known to me. The only social black and yellow, ground-nesting bee I could imagine was a bumble bee, but everyone knows what they are. Instead, the image of yellow jackets sprang to mind. They fit the description, and I knew something about them based on having been stung when mowing my lawn.

"It sounds to me," I said delicately, so as not to seem overbearing, "as if these are not bees but more likely yellow jackets."

"Whatever! How do you kill them?"

Now I was trapped because I don't actually kill such things unless there are extremely good reasons to do so. So in all innocence I began to describe my philosophy of what to do about yellow jackets. "Yellow jackets are beneficial predators of less desirable, even pestiferous, insects . . . ," I began, but before I could say anything else, the fellow blurted out:

"I just dowse the damn things with gasoline and set them on fire." He then walked off to another exhibit. I have become so fond of this technique that when folks now ask me how to control household termites I reply simply, "Dowse the damn things with gasoline and set them on fire."

As you might imagine, it is not easy taking the role as champion for the underdog, but my sympathies distinctly lie with the largely unknown and misunderstood members of the order. Whenever possible, I do my best to defend and promote Hymenoptera to the public, and for good reason. As a group, they are extremely important to our own well-being. So important, in fact, that they have been referred to by one prominent entomologist as "the most beneficial of the insects"

(Waldbauer 1998). Although this may seem like hyperbole to some, the goal of this book is to consider the role that Hymenoptera play in the environment and our daily lives, to see if they really are as important as hyperbole suggests, and to provide an introduction to one of the most intriguing insect groups of them all.

1

The Impact of Bees, Wasps, and Ants on Our Lives

MOST HUMANS WOULD as soon avoid bees, wasps, and some ants as the plague, and not without some cause. Never mind rattlesnakes, scorpions, or spiders—based on numbers alone, Hymenoptera are the most likely poisonous animals to be encountered by humans in their daily lives (Banks and Shipolini 1986). Yet deaths caused by their stings are relatively few, running an average of 44 per year in the United States from 1979 to 1990 (Levick et al. 2000). To provide some perspective, death from dog bites for a similar period ran nearly half that (Centers for Disease Control and Prevention 1997), but we seem not to have the same hysterical reaction to man's best friend as we do to being stung. Death notwithstanding, the psychological baggage attached to being stung can, itself, be severe to the point of morbid obsessiveness (or entomophobia, an actual medical condition) in some people. Although the vast majority of Hymenoptera are non-stinging—and then only in the female of the species—all species apparently are guilty by association: merely mention the word *wasp* and people duck for cover.

Fortunately for humans, and doubly so for Hymenoptera, the latter remain mostly unseen by the former. But this ignorance has a negative side, which is that humans have relatively little idea about what Hymenoptera do for us. Consequently it becomes unrealistic to ask the gardener or the public to care about them based on some sense of ethical responsibility for their well-being when humans scarcely know that Hymenoptera exist and are afraid of the ones they do know about. Therefore, before proceeding onward to the chapters concerning individual groups of Hymenoptera in the second part of this book, I feel it my responsibility to make a case for embracing the group as a whole. To do so we'll first examine the possible economic values of the group in terms of cold, hard cash, and then take a more realistic approach to their value as helpmates in the world we all share.

Economic Value, or a Monetary Approach to Appreciating Hymenoptera

We humans tend to base much of the understanding of our own lives on extrinsic monetary values. It is not readily apparent to most of us that the natural world we live in provides a lot of free services for which we pay nothing, and in the case of Hymenoptera often resent. Although it is difficult to attach monetary value to these services, some recent studies have been published that attempt to evaluate the economic value of natural ecological services provided by wild insects.

According to *Science Daily* (2008) "the worldwide economic value of the pollination service provided by insect pollinators, bees mainly, was [U.S. $217 billion] in 2005 for the main crops that feed the world." This figure did not take into account "the impact of pollination . . . [on] seeds used for planting, which is very important for many vegetable crops as well as forage crops and thereby the whole cattle industry, nonfood crops and, perhaps most importantly, the wild flowers and all the ecosystemic services that the natural flora provides to agriculture and to society as a whole."

Speaking solely of the United States, a study published in *BioScience* by an entomologist at Cornell University and the conservation director at the Xerces Society for Invertebrate Conservation presented evidence that insects in general provide about $57 billion worth of free services per year (Losey and Vaughan 2006). The study also stated that this figure "*is only a fraction* of the value of all the services insects provide" (emphasis mine). Without going into the gruesome details of how these figures were derived, I'd like to point out a few of their conclusions modified in part with particular reference to Hymenoptera.

In the study by Losey and Vaughan, the authors purposefully used the term "wild" insects to remove from consideration the economic value of such factors as domesticated honey bees (pollination, honey, wax), insect products such as shellac or dyes, human-directed biological control (the purposeful use of beneficial insects), and the production of medicines (such as bee antivenins). Nor did they include insect services such as decomposing animal carcasses or plant materials, both extremely important in recycling nutrients and nearly impossible to quantify. These indirect economic values were omitted on the basis that figures have been published for the direct economic benefits and that indirect economic values were too difficult to estimate. For example, the decomposing aspect, to be blunt, was excluded because no one other than gardeners places much value on rotten things (Figure 6).

The overall values presented in this study, then, are extremely conservative and limited to those subject areas humans assume to be of value to themselves, as well as what might conceivably be quantifiable: the benefit of insects to food and field crops, cattle raising, fishing, hunting, and wildlife observation (especially bird watching). The study paid scant attention to the "natural world" that supports everything else we need for survival as a species. That world, of course, is where most of the 150,000 described species of Hymenoptera live. Regardless, much of the data provided in the study is important as a guide to understanding our dependence upon insects if only from a purely selfish view. Once we have some concepts of our own self-centered world, we can move on to a more generous understanding of the world as it really is.

Let's start with the easy stuff first, namely pollination. Few of us realize that insects alone pollinate nearly one-third of all the food we consume. Although many insects act as pollinators, by far the most common and efficient of these are bees, both the solitary sorts and domesticated honey bees and bumble bees (Figure 7). Solitary bees are much more common than one might think. For example, during a study conducted over three years at 15 different sites in Michigan, 166 species of bees were reported (Tuell et al. 2009). These sites consisted of patches of highbush blueberry (*Vaccinium corymbosum*), a native North American crop, and the areas surrounding them. Not all these bees were pollinating blueberries, but the study demonstrated the abundance of native bees in a relatively small, simple ecosystem.

Losey and Vaughan (2006) estimated that, excluding honey bees, the yearly value of "native pollinators—almost exclusively bees—may be . . . almost $3.07 billion of fruits and vegetables produced in the United States" (Figure 8). This is a respectable figure for those lonely, isolated, solitary bees that gardeners almost never think of. But remember this figure does not include honey bees. Figures cited by Losey and Vaughan, based on the work of other scientists, put the yearly U.S. value of honey bee pollination at somewhere between $6.8 billion to $16.4 billion in terms of 2003 dollars. Another source has suggested the amount for all pollination services combined is nearly $40 billion if other secondary commodities such as milk and beef derived from seed-grown pollinated plants (such as alfalfa) are included (Shepherd et al. 2003). Not a bad year's work for a group of insects.

Most gardeners will concede that bees serve a purpose, even if it's only to pollinate their own backyard vegetable plants and fruit trees. Yet other groups of Hymenoptera are equally important in their own way, and two of the most important of these are predatory (Figure 9) and parasitic wasps (Figure 10). In their article Losey and Vaughan did not single out these two groups, but described instead "native beneficial insects." They assigned the figure of $18.77 billion to crop losses in the United States per year and estimated that if there were no native beneficial insects an additional $4.5 billion would be lost. This figure is for all predatory and parasitic insects, not just Hymenoptera. However, we can narrow it down just a bit by some information from another source that reported that predaceous and parasitic Hymenoptera combined account for nearly 60 percent of all insect predation on other insects (LaSalle and Gauld 1991). Thus, we might be generous and conclude (rightly or wrongly) that Hymenoptera account for more than half of the $4.5 billion worth of free services strictly in the realm of U.S. agricultural control.

The final data presented by Losey and Vaughan relate to the value of insects as food for the benefit of animals involved in small game hunting, migratory bird hunting, sport and commercial fishing, and wildlife observation (mostly birding). Figures are given for these human recreational activities because they are somewhat quantifiable in terms of money spent and the amount of insect biomass (a quantity of biological weight) that it takes to feed this wildlife. The total figure derived for one year in the United States was $49 billion in free services to humans by insects consumed by the animals we hunt, eat, and watch. There is no way to assess what proportion of these insects are Hymenoptera, but given the fact that the order is the second largest in terms of numbers of species, it would be a fair guess that Hymenoptera are also a fair share of consumables eaten by the sorts of animals under question. In this respect, ants (Figure 11), which are nowhere mentioned by name, would likely be one of the larger nutritional resources available to other animals because they constitute one of the largest amounts of biomass in the insect world.

There is little doubt that bees, wasps, and ants have economic value for humans even if it is difficult to quantify monetarily. E. O. Wilson, Harvard entomologist and double Pulitzer Prize winner, has called insects "the little things that run the world" (Wilson 1987), and the order Hymenoptera has been referred to as "the most beneficial of the insects" by another prominent entomologist (Waldbauer 1998). Given the huge inventory of insects to chose from, why is it that Hymenoptera might seem to be just a bit more prominent than other groups? I will attempt to answer that in the following discussion.

Hymenoptera as Critical Components of the Natural World

Although monetary concerns are generally of importance to humans, there is an oft-neglected aspect of our lives that we ignore at our own peril, namely the natural world in which we must exist with or without money. There is a huge part of the world that is not human and that should be investigated on its own terms, without thoughts of financial gain.

Pollination, Flowers, and Gardens

Flowering plants rely heavily upon pollinators to set seed, which they surely need to reproduce. As Charles Michener, one of the world's preeminent bee experts, has stated "Pollinating insects are essential to our gardens, to most of the earth's flowering plants, and to human beings who are dependent, like every organism, on the web of life" (foreword, *Pollinator Conservation Handbook*, Shepherd et al. 2003).

Flowering plants have evolved flowers not for the benefit of humans—surely a blow to our fragile egos—but for the attraction of pollinators (Figure 12, 13), whether bird, bat, fly, beetle, or bee. According to Danforth et al. (2004), "The enormous radiation of the flowering plants may be due in part to the nearly simultaneous diversification of the bees." Therefore we may well pay homage to all pollinators, of which bees are the most abundant and arguably the most important, for the very reason many of us garden: the flower. True, humankind has taken the flower to new heights by virtue of our own pollinating and hybridizing efforts, but where would we be if the native flowering plants and their pollinators were not there to begin with? Monetary tallies could probably be found somewhere to account for the production of commercial flowers, nursery stock production, landscape services (both planning and maintenance), garden tools, and garden ornament, all products of the origin of flowers. But nowhere could we find figures of the importance to our own psychological and societal well-being that gardens have provided for millennia. Consider a human world without the flowers that evolved to please insects and a few other beasts, and we would be cultivating gardens of grasses, ferns, mosses, lichens, algae, and conifers.

To drag this dead horse one step further, flowers did not generously evolve merely to attract insects and other animals to a free buffet of nectar and pollen, though that was the inducement. What flowers are really for is their own singularly selfish purpose of reproduction. Plants, therefore, ultimately dupe their pollinators into the job of producing new plants in return for a little nourishment. Of course, plants don't realize that without their own reproduction there would be little for the hoards of other herbivorous (plant-feeding) animals to eat except for grasses, ferns, mosses, lichens, algae, and conifers. Imagine such a simple world, and we can be thankful that pollinators arose to service flowers and in so doing to serve humankind as well.

Biomass, or Food for the Masses

Pollination results in plant reproduction, which is a critically important process for animals that feed directly on plants. Scant attention, however, is paid to the services insects provide by sacrificing their own lives to feed the birds, fish, mammals, reptiles, amphibians, and other life forms (including many people throughout the world) dependent upon them. Researchers have attempted to quantify some aspects of these services ($49 billion per year in the United States), but it would seem to be a feeble amount compared to what goes on in the real world.

Hölldobler and Wilson (1994) estimated that half of the world's living insect tissue, their biomass, is represented by the social insects. Biomass refers to the weight of biological material, given as either a dry or living measure. It is usually cited as a weight per unit area for comparative purposes to indicate the amount of a given resource in a particular environment. In the Hölldobler and Wilson paper (which is slightly outdated), they stated that the social insects represent about 13,500 species of which ants represent about 70 percent (9500 species), bees 7 percent (1000 species), and wasps 6 percent (800). (Termites make up nearly all the remaining social insects at 15 percent, or 2000 species). If one does the math, this means that the social Hymenoptera alone make up 41 percent of the world's insect biomass (that is, 83 percent of 50 percent). And don't forget there are another 95,000 described species of Hymenoptera in addition to the social species, so I think it would be fair to state that the order Hymenoptera probably represents half the biomass of insect life on Earth.

This means that there are large amounts of hymenopteran bodies available to be eaten by all sorts of other animals from lizards and frogs to anteaters and skunks to birds, bears, and even humans. The importance of this biomass to the lives of other creatures would be difficult to measure, but as with the hymenopteran contribution to plant reproduction, it would intuitively appear to be of great importance to the natural order of the world.

Predators and Parasitoids: Natural and Biological Controls

Not only do Hymenoptera provide a huge biomass to the organisms that feed upon them, but they are, themselves, great deprecators of other life forms. Based on the feeding habits of known world Hymenoptera, we can estimate that nearly 89,000 species represent parasitic forms and more than 18,000 species represent predatory forms. I exclude ants from these categories as they are rather difficult to summarize as a group, but there are predatory species of ants as well.

In total, then, approximately 75 percent of all Hymenoptera (excluding ants) are feeding entirely on other insects, with a few attacking noninsect hosts such as spiders, mites, ticks, pseudoscorpions, and even nematodes. This destruction of the lives of others is referred to as natural control, meaning that it goes on in nature without the aid of humans.

The extent of natural control and ecological stability—what most folks call the balance of nature—that Hymenoptera exert on the living world is unimaginable and incalculable. It is the reason that, with few exceptions, insects have not taken over the world more so than they have. Based on lions or wolves, it might be fairly easy to imagine the toll that predators take on their various sorts of prey, but our knowledge of life's complications is depleted with regard to the subject of insect predation or parasitism. For instance, researchers in Germany have discovered that plant-feeding caterpillars avoid areas in which they sense the wingbeat of predatory (and probably parasitic) wasps. In so doing, even the wingbeats of bees gathering nectar and pollen proved to reduce caterpillar feeding damage in controlled experiments by nearly 60 percent (Black 2008). In the natural world all things are likely.

In our human way of thinking, when natural control fails to work, populations of organisms such as insects, mites, weeds, and pathogens go unchecked and problems arise. We are familiar with this in the agricultural or horticultural worlds, and we refer to these organisms as pests when they experience population explosions. In an effort to reduce these pest populations some researchers use a technique called biological control, which is nothing more than an artificial attempt to return an unnatural situation back to a natural one using the organisms that normally keep the pest under control in the real world.

Biological control is a technique dating back thousands of years. All too appropriately, a hymenopteran is credited with the first such case in recorded history. In China, predatory ants were manipulated in mandarin orange trees to counteract foliage-feeding insects. Bamboo poles were interlaced among the trees to act as passageways for the ants. The modern age of science-based biological control might be said to have started at the turn of the nineteenth century, again with citrus trees, but this time in California. Although earlier attempts were made at biological control of organisms, this is considered the first case of the completely successful use by humans of one insect to control another. In this case the hero was a little beetle.

Biological control is still used in present times, although not without its own set of negative aspects, largely related to possible unintended, nontarget effects. Looking at its successes, however, we can draw a few hymenopteran-related conclusions. Greathead (1986) reported 570 parasitoid species (almost exclusively Hymenoptera) that had been released against 274 pest species (Figure 14). Of these parasitoids, 393 species proved effective in completely or satisfactorily suppressing 216 pests and another 52 cases resulted in useful reductions. Drawing a lesson from the use of biological control by humans, we can see that relatively small numbers of parasitoids have been used to "correct nature." If a mere 400 species are effectively correcting nature through our own manipulation, just think what the other nearly 90,000 parasitoid species are doing to control naturally occurring populations of insects—not to mention the 18,000 predatory wasps, about which much less is documented regarding their contribution to natural control.

In addition to the reduction of insect populations and the direct monetary gains received from improved crop production, there are some side benefits that almost no one credits to the use of biological control. For one thing, it reduces the future capital expenditure for pesticides, labor, and specialized equipment no longer needed to protect the crops—in most cases successful biological control returns the crop environment back to the way it was prior to a pest's arrival. Biological control also benefits wildlife through the reduced use of pesticides, a concept first introduced by Rachael Carson (1962). In addition to directly killing wildlife, chemicals in pesticides can be sequestered in their bodies, mimicking or blocking hormones and resulting in reproductive and behavioral abnormalities that adversely affect the host or its progeny (Hoddle 2003).

In truth there is no way to define the Herculean services provided by parasitoids and predators in the natural environment. One thing they do for certain is to maintain structure and stability within a world that could otherwise exist in constant chaos. For those interested in the topic, I recommend my earlier attempt at explaining the roles of insects in the ecological balance of the garden: *Insects and Gardens: In Pursuit of a Garden Ecology* (Grissell 2001).

Nutrient Recycling and Seed Dispersal

Recycling is an oft-used word in today's era of environmental concern. In the garden we recycle tree bark as mulch and other organic materials as compost. In the natural scheme of things, recycling has been carried on for eons by living organisms such as earthworms, bacteria, protozoa, fungi, and insects, which break down organic debris into elemental particles. The components of what was once a tree or a mushroom eventually becomes a daisy, a bear, a cat, a butterfly, or even a gardener.

Among the insect world, social insects compose roughly 40–50 percent of the world's biomass, and of that about 70 percent of species are ants, most of which live in the ground. In some parts of the world ants are the predominant component of insect biomass. One hectare (2.5 acres) of Amazonian rainforest soil, for example, was reported to contain 8 million ants and 1 million termites, with the two taxa accounting for 75 percent of total insect biomass. In Brazil, ants accounted for 30 percent of the arboreal arthropod biomass, exceeding that of terrestrial vertebrates by four times (Hölldobler and Wilson 1990).

Ants are great churners of the soil, making vast underground tunnels, aerating the soil, and moving organic matter from above to below ground. Not only is the organic matter being moved, it is being shredded so that other organisms such as bacteria and fungi can process it further along its route to humus. By providing these services, ants are credited as among the most vital elements of soil structure and nutrient recycling. This in turn, provides the plant community opportunity for optimum development.

In the process of moving all this organic matter from place to place, ants provide yet another unheralded and valuable environmental service. They are seed dispersers, a process referred to as *myrmecochory*. In fact many plants are dependent upon ants for the movement of their seeds from near the parent plant to new ground, where they may develop without competition from their own parents. These plants go by the not entirely melodious term of *myrmecochorus*.

More than 3000 species of plants in nearly 70 families have been reported using ants as seed-transport devices, and this is likely just the tip of the iceberg. In some ecosystems it has been estimated that 40–50 percent of all herbaceous plant species rely on ants for seed dispersal (Gorb and Gorb 2003). Some such plants familiar to gardeners include corydalis, delphinium, bloodroot, trillium, and viola. In many cases the seed has an attractant food structure called an elaiosome that induces an ant to drag the seed back to its nest, strip off the food to feed to its young, and then abandon the seed itself either in or near the nest. Some tropical bees are known to do this as well, but ants are apparently the main stooge in this plant scam.

Antivenins (Antivenoms)

Antivenins are a biological product used in the treatment of venomous bites or stings. The production of antivenins is, of course, only a positive aspect of Hymenoptera relative to the more negative aspect of being stung in the first place—or in any place for that matter. Nevertheless, it is positive if you are in the throes of dying from anaphylactic shock, which is a severe allergic response due to a toxic substance being injected into the body. Anaphylaxis, which can result from a single sting or bite, produces difficulty in breathing and death in a matter of minutes if left untreated. Antivenins are developed from wasp and bee species as an antidote to the sting of those species, and just as with rattlesnake wranglers, there are wasp wranglers who collect wasps for the raw materials from which to produce these life-saving products.

Negative Aspects of Hymenoptera (as Viewed by Humans)

In the spirit of objectivity, I suppose it is necessary to mention that bees, predatory wasps, and ants do occasionally irritate the hell out of humans. Even parasitoids, which can scarcely be thought of as dangerous or scary to humans, will occasionally prove a bit surprising. Rather than discuss ways to control or mitigate the problems caused by these creatures in this section, I've included some methods for discouraging problematic Hymenoptera at the end of chapter 4.

Stings and Bites

Bees and wasps can occasionally be dangerous to one's health. Of all the negative aspects of Hymenoptera, being stung or bitten would have to be the worst. The two actions represent quite opposite aspects of pain, a sting being administered by the rear end of the creature, whereas a bite comes from the front end. In most cases a sting is worse than a bite. In the United States and Canada, not many stinging wasps or bees inflict serious bites; they are mostly concerned with the other end. The tropical stingless bees have turned the other cheek, so to speak, and have little choice other than to bite. Fire ants and harvester ants, on the other hand, generally bite first and then sting with serious consequences. Luckily, in the United States and Canada there are few biting ants of any consequence. The bulldog ant of Australia is about 1.5 inches (4 cm) in length and has mandibles (jaws) longer than their entire heads. Both ends of this ant are painful. According to my colleague Justin Schmidt, the bullet ant of South America is at the apex of Hymenoptera-induced pain. Schmidt, in fact, has published an index called the "Justin O. Schmidt Pain Index" (Schmidt 1990), which is based on his own experiences of being stung by Hymenoptera. The list compares stings somewhat along the lines of fine wines (see chart opposite).

To my knowledge no other insects actually sting using an apparatus housed at the rear end, but certainly many bite and some have stinging (also called urticating) hairs. Some so-called biting insects, such as horseflies and deerflies, really don't bite, they puncture, because their jaws are modified into bloodsucking stilettos, which they jam into a person's skin. These are equally as painful as some of the Hymenoptera listed above, but these biters have to alight and begin a probing process, which often gives the victim enough time to brush them off without much damage. I have been stung by Hymenoptera up to the Schmidt 4.0 level, but nothing has been as painful in my experience as grabbing a saddleback moth caterpillar by accident. These have stinging hairs that break off in the skin and leave a persistent burning effect for hours. Personally I would rate these caterpil-

1.0 **Sweat bee:** Light, ephemeral, almost fruity. A tiny spark has singed a single hair on your arm.

1.2 **Fire ant:** Sharp, sudden, mildly alarming. Like walking across a shag carpet and reaching for the light switch.

1.8 **Bullhorn acacia ant:** A rare, piercing, elevated sort of pain. Someone has fired a staple into your cheek.

2.0 **Bald-faced hornet:** Rich, hearty, slightly crunchy. Similar to getting your hand mashed in a revolving door.

2.0 **Yellow jacket:** Hot and smoky, almost irreverent. Imagine W. C. Fields extinguishing a cigar on your tongue.

2.x **Honey bee and European hornet:** Like a match-head that flips off and burns on your skin.

3.0 **Red harvester ant:** Bold and unrelenting. Somebody is using a drill to excavate your ingrown toenail.

3.0 **Paper wasp:** Caustic and burning, distinctly bitter after-taste. Like spilling a beaker of hydrochloric acid on a paper cut.

4.0 **Pepsis wasp (spider wasp):** Blinding, fierce, shockingly electric. A running hair drier has been dropped into your bubble bath (if you get stung by one you might as well lie down and scream).

4.0+ **Bullet ant:** Pure, intense, brilliant pain. Like fire-walking over flaming charcoal with a 3-inch rusty nail in your heel.

lars as the worst pain of any I've experienced, even including the yellow jacket that once used the inside of my mouth for target practice.

Plant Damage

In the entire order Hymenoptera there are probably somewhere near 10,000 species in which the larvae feed directly on or in plant tissue. Most of these are rather harmless in the scheme of things. The gall formers, for example, induce fascinating and even aesthetically pleasing plant formations, but rarely cause economic or even cosmetic damage to the plants they infest. In truth most galls go unnoticed by the ordinary observer. There are several internal seed feeders that occasionally cause damage when attacking agriculturally important plants such as alfalfa, almonds, apples, soursop, pistachio, and others, but these are in the minority. Occasionally ants will bury the crowns of plants, thus killing them, in their effort to create colonies. In this case the death is accidental, not intentional. Fire ants are disruptive to lawns, especially in areas with sandy soils.

Of all the Hymenoptera, sawflies generally are considered the most economically destructive because they attack a wide variety of plant hosts, often in large numbers. Some examples are common leaf feeders known to gardeners, such as the roseslug, pearslug, and hibiscus sawfly. The group also includes leaf-miners and many species that attack coniferous trees. Larvae of sawflies are often found in large masses, feeding at the tips of branches; they can be devastating in the extreme, especially in forest ecosystems.

Household Nuisances and Structural Damage

The least troublesome household nuisance caused by Hymenoptera is generally ants in the pantry. Although irritating or disgusting to some, the presence of these sugar- or protein-feeding ants is not likely to cause pain or destruction.

An entirely different problem than ants is infestations of the honey bee and social wasps, such as yellow jackets and hornets, that gain entry to the house. Unscreened attics or even breaches in the siding may allow access to hollow, protected areas such as stud walls, in which large colonies may eventually develop. I've heard of both honey bees and yellow jackets penetrating interior sheetrock walls once their colonies have built up large enough to overwhelm the limited space provided by a stud bay.

Other wasps are occasionally found around or on the home. One placid social wasp, called a paper wasp, builds a hanging, downward-facing paper comb under house eves. These wasps are usually so few in numbers and nearly harmless that they can be left on their own with no objection. Occasionally a solitary wasp, such as a mud dauber, organ-pipe wasp, or potter wasp, will build a mud nest on window screens, siding, or under the eaves of a house, but these are mild-mannered, isolated beings that do not live in large aggregations. Solitary wasps rarely inhabit the nest they build; it's merely a home for their larval stage.

Generally a more serious concern than nuisance value is the structural damage to wood caused by carpenter ants and carpenter bees, both of which infest wood, but unlike termites, do not eat it. Carpenter ants normally live in the soil, under rocks, or in rotted tree trunks. They will invade new wood but seem to prefer wood that has some degree of rot as a starting point for their colony. When I had some remodeling done on my house in Maryland, there was an old wooden screened porch that had to be demolished. As it fell to the contractor's hammer, thousands of carpenter ants poured out from its rotten frame. A casual observer might say the ants had chewed up my porch, but probably all they did was take advantage of an already rotten situation. Later, when a new addition was added, I discovered carpenter ants in one of the basement walls. There had been some slight water damage, and the ants apparently took advantage of this moist wood to begin a new colony in new wood. However they become established, carpenter ants are not a good thing to have in the house.

Carpenter bees are solitary, excavating tunnels straight into new or old wood to provide a home for their young. They prefer bare, unpainted wood, and will even burrow into stained or pressure-treated material given a chance. A single female does not cause much in the way of structural damage, but the problem arises after years of repeated infestations by generations of bees. Although the bees are solitary, each nest produces up to ten new bees, of which some are female. These females tend to nest in the same vicinity as the nest from which they arose, so one may reuse the old nest and the rest begin new nests nearby. It does not take a degree in mathematics to see that the population grows exponentially if nothing is done about it. I've read of wooden water towers collapsing after the structural integrity of their staves had been compromised. From personal experience I saw our neighbor's wooden carport turn into a hollowed frame of a structure that collapsed into a pile of sawdust as it was being taken down. This could have been prevented if the original female bee had been stopped or her nest plugged up—or maybe if the carport had been properly painted!

In truth, structural damage from bees and ants (or even termites) is not nearly so much the fault of the creature but of the homeowner who fails to properly maintain his or her domicile.

2

What Are Hymenoptera?

IN MANY RESPECTS this is the most difficult chapter in the book to digest—even for me. First, it deals with what makes a hymenopteran a hymenopteran and not a coleopteran (a beetle). That is, how does Hymenoptera fit among the other insect orders? This is not exactly easy to explain to those who are not entomologists. Then we stumble into the question of how the various main groups of Hymenoptera differ among themselves. For the convenience of discussing these groups, I refer to them as the sawflies (including wood wasps and horntails), true parasitoid wasps, stinging parasitoid wasps, predatory wasps, bees, and ants. These are somewhat artificial groups based on the main biological modes of life, and I will try to explain why these names are used as opposed to the strictly technical groupings currently being used. There is no obviously easy answer to parts of this dilemma, but we'll attempt an approximation of reality. Then will come some remarks on the untidy business of how professionals handle the numerous groupings commonly referred to as families. If I were you I'd be tempted to skip on to the next chapter, which will be far more interesting, but unfortunately I'm not you so I'll get to the business at hand.

What Makes a Wasp (or Bee, or Ant) a Wasp (or Bee, or Ant)?

There is some disagreement concerning the derivation of the name Hymenoptera, a name that is of more than academic interest. According to some it is derived from the Greek, while others claim the

Latin. (Entomology has always been more interesting to me than etymology.) Most resources I've consulted, including the Oxford Dictionary, derive the word from the Greek. There seems to be no disagreement about the *ptera* part of *Hymenoptera*. It means "wing." The difficulty arises with the choice of whether the prefix should refer to the word *hymen*, meaning "membranous," or the word *hymeno*, meaning "the god of marriage." Hymenoptera have two pairs of membranous wings, so the former is certainly true, but some other insect orders (see below) also share this condition. The latter term, *hymeno*, seems more useful to me as it describes one of the defining aspects of the order, namely the union of the fore and hind wings by means of hooks on the hind wing called hamuli. These little hooks overlap the hind edge of the fore wing, thus uniting the two wings as one when necessary. As Hymeno was the Greek god of marriage, this union seems sanctified. The wings can also be uncoupled, but I don't believe the Greeks had a god of divorce, so they went with the god they knew.

So, Hymenoptera are insects with two pairs of membranous wings, which can hook or unhook as needed. This hooking takes place in flight so that both wings act as a single unit. There are a couple of difficult points to consider here, and you are probably ahead of me on this. What about the Hymenoptera that have no wings, of which ants (Formicidae) are the commonest example? In this case the reproductive castes (males and females) have wings, which they abandon after the mating flight. The workers (female caste) that are seen running about on the ground never develop wings. In other cases, such as velvet ants (actually a wasp in the family Mutillidae), the male has wings but the female doesn't. In yet others, such as fig wasps (Agaonidae), the female has wings but the male doesn't. And in some of the parasitic wasps neither sex ever has wings. For most of us, the difficulty of not having wings is a moot point because we will seldom see a wingless hymenopteran, and ants are so easily recognized that we can forgive them for being wingless.

The next consideration about wings is that there are some other insect orders that have two pairs of membranous wings. Among these are the termites (Isoptera); aphids, leafhoppers, and cicadas (Hemiptera); dragonflies and damselflies (Odonata); lacewings (Neuroptera); and a few others uncommon enough not to mention. Although their wings may be membranous, none has wings that hook together. In spite of this unobservable fact, one rarely mistakes any of the above for Hymenoptera, though sometimes ants and termites are the subject of confusion.

For other common insect orders the hooks are missing as well, but it is easier to separate them simply on the condition of the wings as follows: in beetles (Coleoptera), the fore wings are rigid or shell-like; in flies (Diptera), the hind wings are reduced to minute clubs and appear to be absent; in butterflies and moths (Lepidoptera), the wings are covered with scales; in grasshoppers (Orthoptera) and related groups (such as mantids), the fore wings are leathery; and in true bugs (Hemiptera) the fore wings are partially leathery.

There are several other unusual characteristics of Hymenoptera, but they are fairly esoteric. One is related to their peculiar genetics and the other to the construction and function of the egg-laying device, or ovipositor. Needless to say, neither character lends itself to ease of identification.

The Main Divisions within Hymenoptera, and Why We Won't Use Them

The study of Hymenoptera has recently taken a serious turn toward the molecular aspects of genetic classification, and professionals are beginning to realize we have to rethink our long-held views on the order. The historical divisions and arrangements within Hymenoptera are becoming, to put it mildly, somewhat confusing even for professionals. For many years the order Hymenoptera has been described as consisting of two main lineages or groupings, namely the sawflies (suborder Symphyta) and the parasitic wasps, predatory wasps, bees, and ants (suborder Apocrita). The latter group is then divided again, but we'll get to that in a bit. In most popular textbooks these terms are still recognized and are useful in a mechanical sort of way. Current research, however, suggests that this simple way of viewing the order is not entirely correct, and within a few years I suspect the terms *Symphyta* and *Apocrita* will begin to disappear, at least among professionals.

In a book of the sort you are reading, the temptation is great to ignore vexing problems and pretend that life is simple. It isn't. For example, if I tell you that bees are technically just hairy predatory wasps or that ants are just wingless predatory wasps you might take offense. The difficulties in dealing with bees, ants, and predatory wasps are minor compared to the main difficulty in dealing with parasitic wasps and the parasitic mode of living, which crosses all boundaries of classification. In reconciling groups of Hymenoptera, technically they may be grouped one way (by morphological characters), and practically they may be grouped another (by biological characters). I have opted to use the practical approach for a book such as this and to discuss the order based largely on biology. To do so somewhat compromises the technical approach, so I'm including both a technical discussion for those who might enjoy that sort of masochism, as well as a practical approach for those who don't.

For those obsessive people who need to know almost everything about a subject, I've included the table of hymenopteran families, which is a condensed summary of just about everything you might technically want to know about Hymenoptera. The table, itself, is an arrangement of the current technical classification, but columns 4 and 5 stress the practical aspects of biology upon which this book is based.

From the technical point of view, at the primitive base of the order Hymenoptera, representing the earliest evolutionary branch, are the sawflies, horntails, wood wasps, stem sawflies, and parasitic wood wasps. All are plant feeders as larvae, except the parasitic wood wasps. These hymenopterans are the suborder Symphyta, and its members are referred to as symphytans, or sawflies for short (Figure 15). Adult Symphyta are defined by the fact that the abdomen is broadly joined to the thorax and does not articulate with it as compared to the remaining Hymenoptera. (For those of you who can't recall your basic entomology, adult insects are composed of three body segments: the head, the thorax, and the abdomen). Because the abdomen broadly joins the thorax, it is incapable of independent movement, and symphytans appear to be waistless and somewhat tubular or cylindrical in structure, much like pencils. Less obvious is the fact that the hind wing of symphytans (except the parasitic wood wasps, family Orussidae) has three closed cells. A majority of immatures of this group appear much like butterfly or moth larvae in that they have legs, move about freely, and feed externally on plants. Thus, the immature stages of many species are more likely to be seen than those in the remaining Hymenoptera (that is, apocritans), because the latter almost always have legless, concealed larvae. It should be noted, however, that symphytan species that feed internally in galls or stems are legless, and so are less likely to be seen. Only members of one symphytan family, Orussidae, are known to feed on other insects.

The second group of Hymenoptera includes the true parasitoid wasps, predatory wasps (including stinging parasitoids), bees, and ants. This group is the suborder Apocrita and is composed of two further subdivisions to be discussed momentarily. All adult Apocrita (Figure 16) differ from all Symphyta (and all other insects) in that the first segment of the abdomen (the propodeum) is rigidly fused to the thorax, thus appearing to be part of it. The second abdominal segment (the petiole) becomes modified as an articulating joint. In most apocritans, the petiole is generally small, nearly invisible, and works much like a ball and socket, but in some wasps, appropriately known as the thread-waisted wasps, the petiole becomes elongate and is easily seen (Figure 17). Additionally, apocritans have two or fewer basal cells in the hind wing, and the larvae are legless.

When we examine most parasitoid wasps, predatory wasps, bees, and ants, the "abdomen" we see is composed of segments beyond the second (that is, the petiole). This may seem of little interest to the casual reader, but in the great scheme of things, the construction of the abdomen allows members of this group to do some amazing things that symphytans can only dream of. Put in practical, streetwise terms, having an articulated abdomen contributes directly to the ability of an apocritan to freely flail its abdomen about with enough dexterity to lay eggs into or onto a host (in the case of parasitoid and predatory wasps) or defend itself with its stinger (in the case of predatory wasps, bees, and ants). Sawflies cannot do this. Also, unlike symphytan larvae, those of Apocrita are legless (grublike) and are hidden away either inside a host, a provisioned nest cell, a gall, a seed, or a cryptic hiding place. Most all larvae of this group are carnivorous or pollinivorus, but some feed on plant tissue.

Apocrita is currently divided into two groups, the Parasitica and the Aculeata. The former refers to the large grouping of true parasitoid wasps, whereas the latter are sometimes referred to as stinging

Hymenoptera and include bees, ants, and predatory wasps (in which the stinging parasitoids are also included). This division is based on the structure of the ovipositor, which in Parasitica is used for laying eggs, whereas in Aculeata it is modified into a structure called a stinger, which is used both for self-protection and to subdue prey. It is herein that confusion abounds, because many of the primitive predatory wasps have a stinger but possess the parasitic behavior found in the Parasitica. They simply perform their task of parasitizing other insects in a slightly different way. That is why I call them stinging parasitoids and treat them with the true parasitoids instead of the predatory wasps to which they technically belong. The terms Aculeata and Parasitica, therefore, are beginning to fade from use based on the many specialists actively working with Hymenoptera who realize that relationships within the group are much more complicated than was previously thought.

With the knowledge that the definitions of the above three groupings (that is, Symphyta, Apocrita-Parasitica, and Apocrita-Aculeata) are in the process of change as well as being somewhat difficult to comprehend from a practical perspective, I approach the matter of recognizing groups of Hymenoptera from a practical bias. The proposed divisions used are not perfect, and there are some biological intergradations between them, but still they give us a way to group hymenopterans together so that they are more easily discussed. In truth, once one gains a general impression (or gestalt, as taxonomists often say), it's not difficult to recognize most Hymenoptera to the groups given below. The bees, sawflies, predatory wasps, and ants are fairly readily distinguished by their size and the fact that they are generally found visiting flowers or tramping along the ground. The majority of parasitoid wasps are distinguished by the fact that if you don't see it, it likely is one! Here I introduce the five main biological groups, each of which will be discussed in more detail in the second part of this book.

Sawflies, Horntails, and Wood Wasps

These wasps form the numerically smallest group of Hymenoptera at about 5 percent (about 8200 species). All are plant feeders as larvae, except for the parasitic wood wasps (Orussidae), which are parasitic on wood-boring beetle and wasp larvae. Because this rather rare (and never seen) family poses only a minor problem relative to a discussion of Hymenoptera, I include it in the discussion of sawflies rather than with the parasitoids. This group corresponds to the suborder Symphyta. Researchers seem to agree that sawflies, horntails, and wood wasps, based on the broadly attached, nonpivoting abdomen in adults (Figure 18), complex wing veins, caterpillar-like larvae, and plant-feeding immature stages, really do form a natural group.

True Parasitic Wasps and Stinging Parasitic Wasps (Parasitoids)

By far, the majority of Hymenoptera species belong to this group— more than 60 percent (89,000 species). A great many of these wasps are in the microscopic size range and are sometimes referred to as microhymenoptera. They range in size from smaller than the period at the end of this sentence up to 7 inches (17.5 cm) in length if the very long ovipositor is included. There are two types of parasitic wasps, namely the true parasitoids (about 76,000 species; Figure 19) and the stinging parasitoids (about 13,600 species). They differ in the construction of the ovipositor, which in true parasitoids is used to insert eggs into or onto a host without moving the host to a protected space or nest. The stinging parasitoids use the ovipositor to sting their host, but the eggs are laid from a basal opening, and in some cases the wasp moves the host to a protected spot (as do a majority of the predatory wasps). The parasitic mode of behavior is somewhat complicated by the inclusion of seed-feeding and plant-galling species within parasitic families (2 percent, or about 2900 species).

Predatory Wasps (Stinging or Hunting Wasps)

Members of this and the following two categories are sometimes referred to as macrohymenoptera because they are relatively large and easily observed. (Some species of these groups are as small as the microhymenoptera and vice versa, so these are simply relative terms.) About 12 percent of all Hymenoptera (more than 17,000 species) fall into the predatory wasp group. These are also sometimes referred to as stinging or hunting wasps because in the process of preying on other insects they both hunt them down and sting them (Figure 20). These wasps are predators in the sense that most adult females sting their prey, usually paralyzing them, and move them to a species-specific place, such as a twig nest, where the wasps' larvae develop. The ovipositor is used only for stinging the host, the egg is not inserted through it, as in the case of true parasitoids, but is ejected from an opening at the base of the ovipositor. There are variations that do not fit the typical behavior for these wasps and more closely approach those of parasitoids and even bees in one subfamily.

Bees

Bees represent about 13 percent (more than 19,000 species) of known Hymenoptera. They are usually easily recognized by their flower-visiting habits and their generally fuzzy appearance. Most bees (Figure 21, right) have branched hairs on their body, whereas most wasps (Figure 21, left) have simple hairs. But as is generally the case, this rule does not always apply: there are species in both groups that have no hairs—or at least no visible hairs! All bee larvae feed on pollen and nectar (or honey) provided by adults (Figure 22) in a nest of some sort. Not all bees, however, are bee-like in demeanor. Some act in devious ways, as we shall see.

Ants

By species, ants make up about 8 percent of the hymenopterous fauna (about 12,500 species). By sheer volume of numbers and weight, they make up a vastly larger percentage of Hymenoptera than all the rest combined. Gardeners generally have few problem identifying ants based on the fact that they are extremely ant-like (the ants, not the gardeners). It's difficult to mistake an ant for anything else, although there are some predatory wasps that appear to be ants (velvet ants, for example). All ants are social, living in colonies either in the ground, in wood, in plant tissue, or in nests built above ground (Figure 23).

The Families of Hymenoptera

When speaking about Hymenoptera the unit of thought that is perhaps most useful (and frustrating at the same time) is the family (see the table of hymenopteran families). The family concept often sums up the most information about a group in the least amount of time. For example, when speaking of the spider wasp family, Pompilidae, we can discuss more than 4000 species, all of which attack spiders. Unfortunately some families are extremely diverse when it comes to the hosts they use, so sometimes we can say only that they attack a wide range of hosts. Within the Hymenoptera numerous superfamilies, families, and subfamilies are recognized. These categories are an attempt to arrange the nearly 150,000 described species into some semblance of order.

Currently there are 75 recognized families of Hymenoptera in North America, with an additional 14 in the rest of the world, but these groupings are not writ in stone for all time to come. Unfortunately, by the time this book is published, the number of families may rise or fall based on the new findings and interpretations that are constantly forthcoming. In order to gain a basic understanding of Hymenoptera, detailed discussions of each and every family is not really necessary, but we will, of course, examine some of them in the second part of this book. To discuss them all would require several more volumes and lots more patience than either of us have.

As a compromise solution to the problem of too much or too little factual information, I have listed all the world's currently recognized families of Hymenoptera in the table at the back of the book. This also includes the current superfamily structure that is used to organize them. In this list are some facts about numbers of species as well as the basic biological habit of each family. The table gives a technical overview of the family names in use as well as some idea of the size of each group in terms of numbers. For some families found in the United States and Canada, I also include subfamily names and data as a means of exploring the diversity found within that family. There are a few families, however, such as Ichneumonidae (35 subfamilies), Braconidae (29 subfamilies), and Pteromalidae (39 subfamilies) for which I have simply given a general summary. There is scarcely any sense in delving deeply into the bowels of those families as they all perform about the same function in the ecosystem, and in many cases researchers are not entirely certain how to define the limits of the groupings to anyone's satisfaction.

Finally, a word about common names. Although some families of Hymenoptera have a single recognizable common name, such as ants (Formicidae), gall wasps (Cynipidae), velvet ants (Mutillidae), spider wasps (Pompilidae), or leafcutter bees (Megachilidae), many do not. Many families are simply referred to by a truncated form of their family name. A few such as rhopalosomatids (Rhopalosomatidae), sclerogibbids (Sclerogibbidae), and proctotrupids (Proctotrupidae) do not trip lightly off the tongue and are scarcely any more pronounceable than the scientific name. And be very, very thankful that the bradynobaenids (Bradynobaenidae) are a rare group you will never have to speak about, at least in polite company. At least one family name has so many recognized common names as to be somewhat useless. When speaking of the family Vespidae, we may be referring to paper wasps, yellow jackets, hornets, potter wasps, masarid wasps, polybiine wasps, and a few others.

To put it bluntly, it is sometimes difficult to talk about some families of Hymenoptera because the family concepts are changing relatively quickly and not everyone agrees on what names to call what groups. Actually, many hymenopterists, of which I'm one, tend to think in generalities that other hymenopterists accept, but which we know to be nearly nonsensical! It has been my pleasure to work with one such group of parasitic wasps that continue to defy all attempts to beat some sense into them. The group is the superfamily Chalcidoidea, which currently consists of more than 20 families and 20,000 described species. At different times, all members of the superfamily have been consigned to a single family or sometimes to as many as 22 families. Researchers are approaching this group from many different perspectives, including molecular analysis of DNA, genetic study of chromosomes, and comparative morphological studies: all with amazingly little success.

As an example of one major problem, we know that within the superfamily Chalcidoidea the family historically called Pteromalidae, with 39 subfamilies, is actually an aggregation of genera and species, some of which may belong in 10 different other families, but we don't yet know what these families might be, whether we should name some new families to solve the situation, or even if we should simply combine all the families in Chalcidoidea into one (as was once done). So we just talk about the family Pteromalidae as if it actually existed. We hymenopterists are very good at ignoring reality in favor of some sort of perceived clarity.

Not all families pose philosophical debates about the limits that define them. For example, if we speak of the family Formicidae, we are covering the entire category known as ants to the general public throughout the world. There are not many such easily categorized groups. The family Pompilidae, or spider wasps, would also fit the bill except that these are rather cryptic wasps hardly anyone has seen, much less that they form a nice, neat, cohesive group. Hymenopterists agree (more or less) that: "Once a spider wasp, always a spider wasp." There are several smaller families of Hymenoptera that pose few problems as well, but they are essentially unknown to the general public.

Once past the smaller, less well known groups, however, the manner in which we group and name Hymenoptera can become confusing even to hymenopterists. For example, there have been seven families of bees and one family of predatory wasp called Sphecidae (or digger wasps). The bees and these wasps have always been treated as lineages distinct from one another, and each was even included in its own superfamily group. That is, both groups were thought to be distantly related, but perhaps with a common ancestor. Everyone knew that a bee was a bee and a digger wasp was a digger wasp, didn't they? It is now evident, however, that bees are, to put it simply, specialized wasps. They arose from within the wasp lineage, not from an ancestor common to both groups. As a result of this finding, hymenopterists now split up what was once the single family Sphecidae into four families in order to place bees into a scheme of classification that seems to fit the facts better than the previous hypothesis.

But the bee plot thickens. Resolving the bee/wasp problem now poses a bee/bee problem. Currently, bees are divided into seven families worldwide, but some researchers suggest that all bees should be recognized as a single family. This debate is rather more academic than we need to pursue, but it simply points out that some aspects of Hymenoptera classification are currently unresolved and may remain so for some time.

Speaking as a systematist (sometimes referred to as a taxonomist), one who studies aspects of biological diversity and nomenclature, it is often not easy to assign insects to precise categories because there are so many species and their morphological, behavioral, and genetic differences frequently tend to overlap or intergrade. Often the best we can do is estimate degrees of relationship and/or distinctness and assign them to hypothetical groupings as information becomes available. As anyone who gardens is well aware, botanists are often found transferring species from one genus to another, changing generic names, or even changing family names, for somewhat obscure reasons. Sometimes it is simply bookkeeping to establish the correct (that is, chronological) historical name. But as often as not, it is because the evolutionary relationships of groups of species, genera, and families of plants are obscure and difficult to reconcile. This obscurity arises because organisms, whether plants or animals, are constantly evolving over time and space and are not precisely fixed genetic statues. Every individual organism is prone to variation, which the taxonomist attempts to codify, but the truth is that organisms have no particular regard for those of us who attempt to study them.

Summary and Resolution of Some Weighty Problems

To resolve the problems inherent in what names to call what groups in Hymenoptera, I have adopted the expediency of using the most recently proposed system recognized in the excellent paper written by my colleague Dr. Michael Sharkey at the University of Kentucky, entitled, "Phylogeny and Classification of Hymenoptera" (2007). These names are used in the table of hymenopteran families. In this way I absolve myself of all problems related to naming things and easily shuffle the blame to my one-time friend. Even he was forced to conclude: "In the course of putting the 'best guess' [of their relationships] together I changed my mind often" (Sharkey 2007). This is by no means a reflection of Dr. Sharkey's talents, though it may be. It is merely an admission that we hymenopterists have yet to resolve some difficult questions concerning the way we perceive hymenopteran classification.

3

Some Basics of
Hymenopteran Biology

IN THIS CHAPTER the various biological wonders of the order are presented for your amazement. We will examine the four life stages in general, discuss the various feeding methods, investigate the stimulating subject of sex, follow-up with some information on the ovipositor (that is, egg-laying device), and finally discuss the various solitary and social mores of the entire group.

One reason for giving a general biological overview of the Hymenoptera is that much of the information discussed below applies to some member or other of the entire order, and it would become overly repetitive if discussed for each of the five groups. Alternatively some subjects apply to some, but not all, members in each of several groups. For example sociality occurs in some members of the predatory wasps, some bees, and all ants; some form of parasitoid behavior occurs in the sawflies, most (but oddly enough not all!) parasitoids, some predatory wasps, some bees, and some ants; and the ability to sting occurs only in the predatory wasps, bees, and ants, but not the sawflies or parasitoids (some can poke, by the way, but not sting). So, we will cover the abundance of general biological information once here, and then to refine some of the details when we get to the specific groups covered in the second part of this book.

Life Stages

All Hymenoptera are holometabolus, that is, progressing through four life stages: egg, larva, pupa, and adult. This is also referred to as having a complete metamorphosis. In this respect Hymenoptera are similar to butterflies and moths, the principle difference being that hymenopteran larvae are for the most part cryptic and hidden away from public view. Within these four basic stages, the lives of Hymenoptera take on a multitude of strange, bewildering, and wondrous forms, many of which are peculiar to the order.

Egg

You might think that an egg by any other name would still be an egg. In Hymenoptera this is not necessarily so. In broad terms, placement of a hymenopteran egg must be where the larva is eventually to develop. The egg may be placed on leaves if the larva is an external plant feeder; inside plant tissue if the larva is to feed in stems, seeds, or galls; on or in a host insect if the larva is a parasite or predator; or on a pollen ball prepared by the adult female if the larva is a pollen feeder. Many of these basic practices will be discussed in the second part of this book, but a few of the more ingenious twists that deviate from the norm are discussed here.

In the case of trigonalid wasps (Trigonalidae; Figure 24), for example, large numbers of eggs are laid on plant foliage. These eggs have a thick, tough shell. When an egg is ingested by a plant-feeding larva such as that of a sawfly or lepidopterous caterpillar feeding on the leaves, it hatches inside the presumed host. But the trigonalid larva does not eat the caterpillar, instead it will only feed on the larva of another parasite already inside the caterpillar. Generally the parasite is a tachinid fly (Tachinidae) or an ichneumonid wasp (Ichneumonidae). If no parasite is present, then the trigonalid larva waits for the host to be parasitized. A trigonalid egg dies if it is not eaten by a caterpillar, and the ensuing larva dies if the caterpillar is never parasitized by another insect. Another species of this family must have the egg ingested by a caterpillar that is then captured, chewed up, and taken back to the nest of a social wasp (Vespidae), where the female regurgitates her haul, containing the unhatched egg, to her awaiting larva. The trigonalid larva hatches, then feasts inside the vespid larva. This is possibly the only recorded case in history where the dinner eats the diner!

A similar case among hymenopterans occurs when the female of a eucharitid wasp (Eucharitidae) lays huge quantities of eggs—up to

10,000 by some reports—on plant tissue. The eggs hatch and the tiny larvae (planidia) are able to move about for a short distance. When an insect wanders by, the eucharitid planidium latches on to it with its jaws, whereupon it receives a free ride. Unless the hosting insect is an ant, the planidium will die because what it really needs is a free ride back to an ant nest. When it finds itself near the larval stage of an ant, it jumps ship and latches onto its dinner. The eucharitid does nothing until the ant larva pupates, then it begins to feed. There is evidence that if a planidium initially hitches a free ride on an insect associated with ants, it might be able to grab onto an ant eventually. This relationship has been shown to work with thrips, but little else is known about the ant-aholics of the family Eucharitidae.

One more interesting note about hymenopteran eggs deals with the production of more than one individual from a single egg. This uncommon feat is called polyembryony and it occurs in four families of Hymenoptera. It is what happens in humans when twins are born, but wasps take the trick just a bit further—in some cases 2000 to 3000 times further. In several genera of the family Encyrtidae (encyrtid wasps), a female lays one or more eggs into the egg of a host caterpillar. As the caterpillar develops, thousands of wasp embryos emerge from the single tiny egg that is less than the size of a pinhead. Of these larvae, a small number serve only to hunt down and kill larvae of other parasites as well as larvae of its own kind produced by females unrelated to themselves who might have also laid an egg in the host. These hunters eventually die, and the remaining larger number of parasites kill and emerge from the host (Alvarez 1997). Imagine a chicken egg producing 2000 chicks, and you have some idea of how phenomenal an event this is.

Larva

With relatively few exceptions (sawflies and relatives), the larvae of Hymenoptera do not move from the point of egg-hatch, depending for survival upon their mother to do the right thing—or at least to do the thing right. The female must lay her eggs exactly where they need to be, or the hatching larvae will be in serious trouble. In the case of social wasps and bees, for example, the larvae are housed perfectly in cells in which the adult female places food for her young to eat.

Because larvae are fed as they develop, this is called progressive provisioning. In the case of solitary bees, the female builds a single cell (or cluster of cells) in which pollen and an egg are placed and then sealed, and the ensuing larva must survive on what it was given. The same is true of solitary wasps, except they usually provision with a precisely chosen range of insect prey depending upon the wasp species. And for parasites, the female lays her egg on or inside a host so the larva has no choice as to where or what its next meal will be.

As noted above in the case of eucharitid wasps, some hymenopteran larvae are slightly self-mobile, relying on the kindness of passing strangers to ultimately arrive at their dining destination. This is an uncommon practice, though, occurring in only three parasite families (Eucharitidae, Perilampidae, Ichneumonidae). The only group of Hymenoptera that can be said to be truly mobile as larvae are some of the sawflies (and relatives), which number but a few thousand species. The majority of larvae of this group resemble butterfly or moth caterpillars; they move freely about, and feed externally on plant tissue. Some such as the wood wasps (Xiphydriidae), horntails (Siricidae), and stem sawflies (Cephidae) are borers, moving about only within the confines of a plant stem.

The length of the larval stage of a hymenopteran is species and temperature dependent. Some species, especially parasites and early-spring gall formers, pass the immature stage in less than a week before emerging as adults. Others may take months. In the case of late-spring gall formers, the larval stage may last almost a year until early the next spring. Many Hymenoptera will overwinter as larvae (or prepupae, see below), thus taking at least several seasons to emerge as adults. Some internal seed feeders are known to take at least a year to develop, and in some cases will remain dormant for two or even three years within the seed.

In general, hymenopterous larvae can mostly be thought of as reclusive, elusive, nonmobile, and helpless if left to feed on their own. They are truly mother-dependent for their ultimate survival.

Pupa

The pupal stage is a quiet period of change from larvalhood to adulthood. Some species go through a prepupal stage, in which larvae become suspended in time between a true larva and a true pupa. This stage often takes place in species that overwinter in immature rather than adult stages. For many species of Hymenoptera, pupation simply takes place in the host or a protected place, without need for a cover (that is, a cocoon), but in some species pupation occurs in a silken or papery cocoon either in the ground, on or near the host from which they emerged, in the cell in which they developed, or even suspended in air from a silken thread. Exact methods of pupation vary even within families of Hymenoptera, so it is difficult to make general statements that are correct in all cases.

Adult

Adults are the final chapter in the life of insects. Unlike humans, who creep slowly from childhood to senility, a hymenopteran springs forth fully formed from pupalhood. Though its juvenile stage may have varied from a few days, weeks, months, or even years, an adult's short life is basically dedicated to reproduction followed by oblivion. The only exceptions are adult queens of some of the social groups of ants, wasps, and bees. For the most part, adults, even the predatory sorts that hunt for the sake of their children-to-be, are consumers of nectar, pollen, and honeydew (the exudates of sap-sucking insects). Water may be equally as important a part of an adult hymenopteran's diet and is readily available as a component of nectar. It is likely that the social wasps (Vespidae), which masticate and regurgitate prey to their young, partake of nutrients, but it appears to be only a side effect of getting the prey to its target. Even adult parasitoids may extract bodily fluids from their host. But perhaps oddest of all are some sawflies (Tenthredinidae) in which the adults can be predatory, whereas their young are plant feeders.

Adult male and female Hymenoptera of the same species generally appear much the same to the naked eye. Sometimes there are color differences that help distinguish the sexes. But in some instances the sexes are completely dimorphic, which means they do not appear similar at all. These differences can take several forms depending upon the species. In some cases one sex has wings (Figure 25, right), but the other is wingless (Figure 25, left). In others, the males are both wingless and grotesquely monstrous (for example, fig wasps, Agaonidae). Some species have individuals with complete wings, stubby wings, or no wings. In one extreme case (Eulophidae) adults appear in four different forms (two kinds of males, two of females) depending upon what part of the host (blood or tissue) they ate as larvae. In some of the alternating generations of gall formers (see Sex, below) one generation of adults is so different from another that historically they were identified as different genera.

Feeding Strategies

As we have just seen, most of a hymenopteran's life revolves around its larval feeding stage, but it is the mother wasp that sets her progeny off in the right direction. She must determine the correct food for her young. It would do no good, after all, to offer up a sumptuous pollen ball if the larva can only survive on flies. In this section we'll attempt to sort out the primary categories of larval feeding that occur among the huge number of species found in the Hymenoptera.

The major recognized categories of Hymenoptera (that is, sawflies, parasitoid wasps, predatory wasps, bees, and ants) do not necessarily follow rigid feeding patterns. They come close, but generally there is always some renegade group that causes an "all" statement to be false. For example, all sawfly species are plant feeders except a very few that are parasitoids, whereas all parasitoid wasps attack other insects except for a relative few species that attack plants. Most predatory wasps (excluding stinging parasitoids) are predatory, but a few species (subfamily Masarinae) collect pollen as food for their young. The categories below, then, pertain primarily to the manner of feeding and not necessarily to a particular group of Hymenoptera. We needn't dwell on specifics until the second half of the book.

For those who find an absence of neat categorization a bit disturbing, remember we are dealing with living entities that care not that we can't find perfectly convenient boxes in which to place them. Additionally, the slippery slope from plant-feeding behavior to that of parasitoids, from parasitoids to predators, and from predators to pollen feeders and beyond is a good indicator that these groups do not remain static but are continually evolving ever-increasing ways to use the resources present in any biologically complex environment (Figure 26).

Plant Feeders

Insects that feed on plants are said to be phytophagous. Among Hymenoptera, direct feeding on plant tissue is the least common method of larval development, but it is found sporadically throughout the entire order. There are several ways of feeding on plants, of which leaf feeding, stem boring, leaf mining, and gall forming occur in the more primitive sawflies, horntails, and wood wasps. Within the so-called parasitoid group, plant gallers are found in one family (gall wasps, Cynipidae; Figure 27) and internal seed feeders (Figure 28) are found in several families of chalcidoid wasps. Feeding from tended fungus gardens is found in some species of ants (Formicidae), and pollen feeding occurs in all bee families and a couple of predatory wasp species (for example, some members of the Masarinae and Vespidae). It is the basic thought among hymenopterists that members of the order evolved from plant-feeding ancestors (symphytans) that gave rise to parasitic and predatory forms, so that the tendency for plant feeding is found rarely and irregularly throughout the evolutionary development of the group.

Predators

In animal terminology, predators attack, kill, and eat their hosts. Hymenoptera, however, generally have to do things differently. It is true that social wasps (Vespidae) and some ant species (Formicidae) attack and kill their host, then chew them up. The ultimate goal, however, is to get the remains back to the nest, where they are regurgitated to feed the young. This is done progressively as the immature stages develop. In this respect, Hymenoptera behave much like birds feeding their young.

The solitary wasps, however, which make up the vast majority of predatory Hymenoptera, act in a distinctly different manner. In the majority of cases a female attacks and paralyzes its host, which she then sequesters somewhere in an isolated crevice, burrow, cell, or structure such as a mud nest. In some cases there is but a single host (for example, a grasshopper), but in others there may be many small hosts (such as weevils). The live, but paralyzed prey, is then consumed slowly as the wasp larva develops. The species of host and quantity given per larva are generally specific to the mother wasp, and there are many variations upon the predatory theme (Figure 29).

Parasites or Parasitoids

Next we come to the most complicated of the feeding types, one that has many twists and turns in its development. When speaking of Hymenoptera, technically the term *parasitoid* is preferred to *parasite* because it defines a modification of the typical parasite behavior. A parasite is an organism that lives on or in the body of its host, but does not kill it—or if it does the death is generally indirect as from a transmitted pathogen. A true parasite is so much smaller than its host that the feeding poses little direct threat. Examples of true parasites are fleas, ticks, and lice. A parasitoid, on the other hand, always kills its host and is generally about the same size, though there are exceptions when more than one parasitoid is consuming the same host. In general, entomologists tend to refer to wasps as *parasitoids* (noun), but we refer to them as *parasitic* (adjective) because the term *parasitoidic* is a mouthful. In this book, then, the terms *parasitic wasp* and *parasitic behavior* are used somewhat interchangeably when we're really referring to parasitoids.

Among insects, Hymenoptera have developed the parasitic life form about as far as it can go. Within all orders they far outnumber and outperform their distant competition, the flies (Diptera), in the abundance of parasitoids, and many specialized terms have been coined to cover the many behavioral variants. The generalized behavior is that a female parasitoid lays an egg or eggs on or in a host, but does not kill it because it must live long enough to support the wasp's developing larva(e). The host invariably dies as the larval wasp(s) mature, but it may take some time. The type of host and even its life stage attacked is highly specific to the parasitoid species and may be an insect egg, a larva, a pupa, or even rarely an adult insect. For example, some parasitoids only lay an egg or eggs within an aphid, another only within a moth larvae in a leaf mine, and some even swim underwater to lay eggs in those of damselflies and drag-onflies. The variants of parasitoid behavior are so specific that they frequently define the species or genus in which they are found.

Ectoparasitoid

Ectoparasitoidism is perhaps the most common larval behavior among parasitoids. The female wasp deposits an egg (or eggs) on the surface of the host and the wasp larva feeds externally with respect to its host. The host is rarely free-living, and more often it is pro-tected within a stem, mud cell, gall, cocoon, leaf roll, leaf mine, or pupal case. The wasp larva is still considered to be external relative to the host upon which it is feeding.

Ectoparasitoid/Predator

In biology there are almost always intergradations of behavior, and the ectoparasitoid/predator is no exception. In this case the female wasp places her egg just as a typical ectoparasitoid would, the egg hatches and eats the host, but then the parasitoid larva wanders off and eats other nearby hosts. This is somewhat common when mul-tiple eggs of a host are clustered in a capsule (think cockroach, spider, or mantid egg case) and the parasitoid can easily move from one host to another, eating as it goes. It also happens sometimes in the case of gregarious parasitoids (see below). If a number of parasitoids are en-closed in a single space munching away on their host, then an incom-ing parasitoid of a different species (a hyperparasitoid in this case) may simply devour as many of the inhabitants as it can eat.

Endoparasitoid

Endoparasitoids are common, but less so than ectoparasitoids. The female wasp deposits her egg(s) on or inside a host. If it is laid on the host, the emerging wasp larva burrows into the host to complete de-velopment. The endoparasitoid larva feeds inside its host, which can be an egg, larva, pupa, or adult, without regard for where the host is located. For example, an internal parasitoid will be inside a host egg or larva even if the host is inside a gall or stem (Figures 30, 31).

Parasitoid Sac

Some endoparasitic wasps form a parasitoid sac, a highly specialized condition in which a larva is feeding inside a host, but part of its body is extruded from the host in a shell-like sac. This is rare, occurring only in dryinid wasps (Dryinidae) that use leafhoppers as hosts. Rhopalosomatid wasp larvae (Rhopalosomatidae) are somewhat similar in that when mature they appear to be housed in a sac-shaped pupal structure that is attached externally to the abdomen of crickets. These are ectoparasitoids, however, and do not begin life inside the host.

Idiobiont or Koinobiont

How the adult female subdues her host so that she can lay eggs and her larvae can feed is an important distinction. In the case of an idiobiont, the female paralyzes the host completely so that it does not move at all. In the case of koinobionts, the female temporarily paralyzes her host and lays an egg, and the host wakes up and wanders off merrily to live its life while the wasp larvae are eating it alive, usually from the inside. A very common example of this type of parasitism is the tomato hornworm—bane of tomato growers—that one often sees covered in white, cottony masses. These masses are cocoons (not eggs as many think) of braconid wasp larvae (Braconidae) that have eaten the hornworm from the inside. The hornworm grows normally, never realizing that it is eating for itself and dozens of wasps inside it.

Hyperparasitoids

When a host already has a parasitoid living in it, some wasps only attack that parasitoid but not the actual host itself. These are primary hyperparasitoids. There are versions of this habit called secondary and tertiary hyperparasitoids in which each level of wasp attacked is attacked by yet another wasp. This may be thought of like a set of Russian nesting dolls. Although primary hyperparasitoids are somewhat common, the other sorts are not.

Gregarious Parasitoids

In this case more than one parasitoid feeds and emerges from a single host. In most cases it is because the female wasp lays more than one egg, while in other cases it is because a single host is attacked by several different females of the same species. In rare cases gregarious parasitoid broods are produced by a single egg in which polyembryony has taken place (as discussed in the Egg section above).

Cleptoparasitoid

Cleptoparasitoid is a term meaning "parasitism by theft." It is used for a form of parasitism in which the perpetrator doesn't attack a host directly, but eats the food provided for it, such as the pollen in a bee cell or the prey item(s) in a predatory wasp cell. The cleptoparasitoid bee or wasp larva thus outcompetes the rightful larva and starves it to death. There are a few variations of this habit, which complicate matters a bit. In some cases the cleptoparasitoid adult kills the host egg, thus allowing her offspring total control of the food source. In another case the cleptoparasitoid larva kills the host egg or larva. This habit occurs in several bee families and several predatory wasp families such as spider wasps (Pompilidae), cuckoo wasps (Chrysididae), and velvet ants (Mutillidae).

Social (or Brood) Parasitoid

Social (or brood) parasitoids use a slight modification of cleptoparasitism that takes place in social Hymenoptera (for example, some bumble bees, vespid wasps, and ants). The solitary female of the parasitoid species invades the colony of a different species, killing its queen. The workers then begin raising larvae of the invading queen, who is the colony's only egg layer. This is much like a cuckoo or cowbird that lays its eggs in the nest of a different species.

Inquilines

An *inquiline* is a species of animal that lives within the nest or abode of another animal, with neither being harmed. In the case of Hymenoptera, an inquiline is basically a freeloader that lives off surplus materials produced by its chosen host. Some gall wasps, for example, create a gall and live happily within it, while at the same time an inquiline gall wasp of a different genus may be living in the tissue surrounding the gall maker. This is an obligate relationship, with no harm to either. The inquiline wasp cannot make its own gall, nor can it live in the gall made by a fly, for example. It must live in a specific gall made by a specific gall maker.

Aquatic Hymenoptera

It may come as a surprise to discover that 150 species of Hymenoptera are considered to be aquatic, living part of their lives in freshwater lakes, ponds, and streams (Bennett 2008). Bennett defined aquatic insects to include "species in which female adults enter the water to search for hosts, those with endoparasitoid larvae inside aquatic larval hosts (even if oviposition is terrestrial); and those in which freshly eclosed [emerged] adults must travel to the water's surface following pupation (even if they develop inside stems of emergent vegetation)." He stated that all such Hymenoptera species are parasitoids, but I'd place one of them, a spider wasp in the family Pompilidae, in the predatory wasp group. According to Bennett, at least 10 different families of parasitoids are known to attack hosts living in water, including dragonfly eggs (Odonata), beetle eggs (Coleoptera), fly larvae (Diptera), moth larvae (Lepidoptera), caddisfly larvae (Trichoptera), and true bug eggs (Hemiptera). The predatory wasps also attack aquatic spiders, though they move the prey to land in order to lay the eggs.

Sex

Hymenoptera, as a group, have several somewhat unique and surprising developments with respect to their sex lives and genetics. If we were to couch hymenopteran reproduction in human terms, it would go something like this. A female mates once, or perhaps several times with different males, and stores the collected sperm in a special internal chamber. She then divorces her husband or possibly shoots him as a matter of expediency because he is of no further use. From that point forward, the female decides when to produce a baby and which sex it is to be. She would continue having sex-specific offspring the rest of her life with no man in sight. A less common variation of this sex life is that of a female who could produce children without ever mating at all. In some cases they would be all males, in some cases all females, and in some cases there might be both female and male progeny.

Returning to Hymenoptera, the first scenario given above is standard practice for the majority of the group. Two necessary factors are involved. First, the female must have a sperm storage unit, which is called a spermatheca. This is common to all insects, by the way, not specific to Hymenoptera. Second, there must be some way for the female to regulate the sex of her eggs as they are being laid, of which there are several variants. If the female allows sperm to exit the spermatheca as an egg is being laid, it is fertilized and has a half a set of chromosomes from the mother and half from the father (that is, the offspring are diploid, having a full set of chromosomes). This egg develops into a daughter. If she closes the spermatheca, however, the result is an unfertilized egg with a half a set of chromosomes derived only from the mother (that is, haploid, having half a set of chromosomes). This egg develops into a son. The latter can also occur in females who have not mated because they have no sperm to release. This system of reproduction is called haplodiploidy and is believed to occur in all Hymenoptera but only rarely in a few other groups, including beetles (Coleoptera), thrips (Thysanoptera), and some aphids and scales (Hemiptera). As a result of haplodiploidy, female offspring have a mother and father, but male offspring have a mother but no father.

Haplodiploidy is a specialized form of parthenogenesis, which is the production of offspring without fertilization, or in other words, virgin births. As we have just seen, Hymenoptera can produce progeny whether or not they are mated. In some species of Hymenoptera, however, unmated adult females can give rise not only to male offspring but also to female offspring. In many respects Hymenoptera are not simple creatures, so it should come as no surprise that some species have found a way around the unmated-all-male conundrum. The basic explanation is that when the nucleus of the diploid egg undergoes reduction division (meiosis, a doubling in number) in preparation to receive the male half of the chromosomes (which are not going to arrive), the doubled chromosome strands fuse back together, resulting in a diploid egg with only female genetic material.

If one examines the hymenopteran system of reproduction a bit closer, several interesting observations come to the forefront, not the least of which is that the consequences of this system are terribly complicated. Haplodiploidy appears to be what has led directly to the independent evolution of sociality in ants, bees, and wasps on almost a dozen separate occasions, yet in other insects that do not show haplodiploidy true sociality has only arisen once, in the termites. We'll revisit the topic of sociality below.

Another result of haplodiploidy is a fascinating event called alternating generations in which one generation of wasps consists of both females and males (the bisexual generation), whereas the subsequent generation is entirely female (oddly enough, the asexual generation). Within insects this alternation of generations is found only in a few aphids (Hemiptera) and in some species of gall-forming wasps (Cynipidae) associated with oaks. In general, the overwintering form of such gall wasps is entirely female (asexual), and the wasps spend the winter encased within the gall they formed the preceding spring. Upon emerging the following spring, the female generation lays both female and male eggs into freshly sprouting plant tissue. The resultant galls mature in only a week or so, and when the male and female adult wasps emerge (bisexual generation) they mate but produce only female progeny (asexual generation).

It might come as a revelation to some readers that a great deal of knowledge about human sexuality was inspired by the study of sex in Hymenoptera, particularly gall wasps. Dr. Alfred Charles Kinsey, founder of what eventually became the Kinsey Institute for Research in Sex, Gender, and Reproduction, began his career as an entomologist at Indiana University. Kinsey studied at Harvard University under the supervision of William Morton Wheeler, a famous ant researcher of the time. Kinsey chose not to study ants, but instead investigated the biology, taxonomy, and evolution of gall wasps. He received his degree in 1919 and published articles and books on gall wasps until he turned his attention to human sexuality in the mid-1930s. According to Wikipedia (2008; http://en.wikipedia.org/wiki/Alfred_Kinsey) "It is likely that Kinsey's study of the variations in mating practices among gall wasps led him to wonder how widely varied sexual practices among humans were." As a direct result of his studies on humans Kinsey produced two books, *Sexual Behavior in the Human Male* (1948) and *Sexual Behavior in the Human Female* (1953), which were received not without some degree of notoriety. And to think that it all happened as a result of the lowly gall wasp.

Mating

For the most part, mating among Hymenoptera, when it does take place, is not much different from mating in other insects. Basically the male searches out a female and takes care of business as efficiently as possible. If he is efficient, he may search out numerous females. In truth, the only thing a male hymenopteran is good for is mating.

In the case of many solitary predatory wasps and bees, males emerge from their nests earlier than females (called protandry). They are then in a position to find and seduce a female immediately when she emerges. In this regard males may become understandably territorial. Each will patrol a suitable area, resting from time to time on a designated perch, and pursue any insect that breeches his domain, especially males of his own species. After all, he is attempting to mate with a female, so he wants no competition from his own species. Because the area in which he emerges is likely to include many nests, a male is likely to encounter a mate sooner rather than later.

Social predatory wasps, bees, and ants have the most complicated mating systems of all Hymenoptera. They are essentially female societies and function well until such time as the colony needs to produce reproductive individuals. In some cases, such as paper wasps and bumble bees, colonies die out in autumn. In late summer, males and new queens are produced and mating takes place. Males die, but the new queens overwinter in protected places. In the case of paper wasps, males may attempt mating on the nest, but it seems they do better by patrolling potential hibernating sites where females are likely to overwinter. Several males may end up keeping vigil on the same site until new queens appear.

According to the University of Florida Book of Insect Records, females of the giant honey bee (*Apis dorsata*, an Asian species typically 1 inch [2.5 cm] in length) are credited among all insects as having the highest number of matings (53) with different males (Cabrera-Mireles 1998), whereas the most spectacular mating behavior is that of the common (European or Italian) honey bee (*Apis mellifera*; Sieglaff 1994). In the common honey bee an aggregation or swarm of males, the drones, cloud the sky awaiting the arrival of a newly emerged queen. The queen is inseminated by up to 20 males, one male at a time. In each case, the male's joy is short-lived as his sexual organ is ripped off (male readers: Be glad you're human!) and remains in the queen until she or the next male removes it and suffers the same fate himself. Thankfully, the male then dies.

Ovipositor, Where Is Thy Sting?

The ovipositor, or egg-laying device, of female Hymenoptera is one of the most perplexing aspects of the group. It is the ovipositor of some wasps, bees, and ants that administers the great, walloping sting so intensely feared by humanity. Although all female Hymenoptera have an ovipositor, not all have it modified into a stinger. It falls to reason, of course, that only females are capable of laying eggs, so that females are the problematic sex when it comes to stinging, and then only females of certain groups of wasps pose the greatest threat. Perhaps we males can take pride in the fact that, for once, we are not involved in the mayhem that threatens the world.

Technically the ovipositor is a highly complex modification and partial fusion of the last two pairs of abdominal segments, each of which has two appendages. It doesn't serve much purpose to go into structural details, but simply put the structure and functionality of these segments divides Hymenoptera into two basic groups. In the primitive, basal group are the sawflies, wood wasps, horntails, and true parasitoid wasps, which include the vast majority of Hymenoptera. In these groups, the abdominal segments and appendages have become elongated and lance-shaped, with two sliding sides and a protective top forming a hollow tube down which an egg passes. The tips of each side lance have serrations that serve as saw blades, cutting a path either into plant tissue or into a host insect. In the true parasitoid forms, a bit of venom passes along the tube and subdues the host long enough to allow egg deposition. Because the ovipositor is flexible in the true parasitoids, an egg can be precisely directed into the exact spot in which it is to develop.

Protecting the lance-shaped ovipositor are sheaths, also modified from the last abdominal segments. These sheaths are actually what is visible when a wasp is not in the process of laying eggs. The ovipositor is released from its sheaths at the time of oviposition, at which point the wasp may appear to have more than a single ovipositor: one embedded in the prey and another (the two sheaths, which may be together or somewhat separated) suspended in midair. In some parasitoids the ovipositor and sheaths are as long as, or much longer than, the female body itself, the record being something on the order of 14 times the body length in a braconid wasp (Townes 1975). The ovipositor of the giant ichneumonid wasp (Ichneumonidae, *Megarhyssa atrata*) may reach 5 inches (12.5 cm) or more in length, which when added to the remainder of the body creates a wasp reaching up to 7 inches (17.5 cm) (based upon my measurements taken at the U. S. National Museum of Natural History). Although these ovipositors appear to be deadly hypodermic syringes to humans—and in some species can penetrate inches of solid wood—they are so flexible as to be useless in self-defense. Among all the true parasitoid species, only a very few with short, stiff, barb-shaped ovipositor sheaths (that is, some ichneumonid wasps, Ichneumonidae) can sting if mishandled. Some stings are painful, and if venom is injected the site may remain irritated for several days. In general, I find it best to not handle wasps with short, pointy things sticking out their rear ends.

The second subdivision of Hymenoptera based on ovipositor structure consists of the predatory wasps (also including some stinging parasitoid forms), bees, and ants. These are often referred to as the "stinging" Hymenoptera and for good reason. In this group, the ovipositor no longer functions as the egg-laying conduit, but has become modified to act as a multipurpose stun-gun/venom injection apparatus. Among solitary predatory wasps, the main function is to subdue and temporarily paralyze prey long enough to preserve it as food for the progeny to consume. Although many such wasps will sting in self-defense, they are not as easily provoked to do so as the social wasps, social bees, and some ants. Among social wasps and most ants (not all female ants sting), the ovipositor functions both to paralyze or kill prey and as a colonial defense mechanism to ward off enemies. In most bees (not all female bees sting), it serves merely as a means of self-protection for the simple reason that no prey is collected and pollen doesn't fight back. Because the ovipositor no longer serves as an egg-laying conduit, the egg is deposited from a genital opening at its base. The female wasp, bee, or ant does not have nearly as precise control of positioning her egg as does a parasitoid. Many species simply plop their eggs into a cell, onto a pollen ball, or on a pile of prey items. Some predatory wasps, however, have a decent aim and place the egg at a precise spot upon one prey item. They do not inject their eggs, as do many parasitoids.

Among wasps, ants, and bees that sting, all but the honey bee (and some of its relatives) are capable and willing of stinging more than once. In fact, yellow jackets and hornets seem to enjoy it almost too much. Honey bees, however, are quite distinct in this respect because when they sting soft-bodied offenders (such as humans) the stinger rips out of their abdomen along with a pulsating sac of venom, and the barbed tip remains imbedded in the skin. The result is unpleasant for humans, and in rare cases can even lead to death, but for the individual honey bee it is certain death. The general advice to victims of honey bee stings is to carefully scrape away the venom sac, but not to crush it, as this just squeezes more venom into the wound.

Unless a person is highly allergic, there are several home remedies to consider when treating bee and wasp stings. I've never tried any of them, but I've read that a poultice made of meat tenderizer (papain) and water is useful. Another poultice is made from baking soda and apple-cider vinegar. As an adult, I've found the use of certain four-letter words distracts my attention from a painful sting, but I generally don't recommend these for children. Advice about stings in general seems to vary widely, and the best thing to do if you are allergic to bee or wasp stings is to follow your doctor's orders and not listen to people like me.

From Solitary to Eusocial Nesting, with a Few Steps Between

By far most adult Hymenoptera are solitary creatures, pairing up only to mate and perpetuate the species. Parthenogenetic species are utterly reclusive, as they don't bother to mate at all. At the other extreme are the colonial or social species that choose to live together in ways not exactly what humans would consider social and so are termed *eusocial* by entomologists. Between the two extremes are some behavioral categories that bridge the apparently wide gap. As with most members of the natural world, however, there are many species of Hymenoptera that fit between the cracks, so to speak, fitting neither one category nor another—or in some cases fitting both! In this section I attempt to explain the main behavioral differences as simply as possible and discuss some exceptions at the end.

Solitary Lives

The vast majority of Hymenoptera lead lonely, self-sufficient, free-living lives. Males live simply to mate and die. Having mated, females lay eggs where their offspring can survive, then they, too, die. Except in very rare instances, there is no direct maternal care, and whether the offspring survive or not is their own problem. This behavior is found in virtually all species of the sawfly and parasitoid groups and is the most common behavior in solitary predatory wasps and bees as well, though many gardeners probably think of these latter two groups mostly in terms of the colonial or social species. The only group known to contain no solitary species is the ants, which are entirely eusocial.

Larvae of the sawfly group simply live on or in plants, whereas those of the parasitoid groups live on or in their host insect. In the case of all solitary bees and the majority of predatory wasps, some sort of sheltered space is used to protect the young. In its simplest form it is merely a crevice between rocks or a space beneath a log. The term *nest* (or burrow) is used when speaking of any hymenopteran that physically excavates or builds a space in which to house their offspring until maturity. The nest can be a simple to many-branched tunnel in the soil or a dead tree that houses one or more cells or chambers, each of which contains a larva and its provisions. Instead of tunneling or boring, some solitary Hymenoptera build structures consisting of chambers, tubes, or pots made from mud, plant scrapings, or even pebbles.

A slight modification of the solitary habit, sometimes called subsocial behavior, is found in some bees and wasps. In this case a female does more than simply provide a nest and food; she progressively feeds and cares for her offspring but departs the nest before they emerge. There is no overlap of generations, no division of labor, and no cooperation between adult individuals.

Sometimes solitary nesting bees or predatory wasps find themselves living in close association merely by coincidence. In such cases a single female bee or wasp finds a suitable nesting site. If the site is large enough, over a period of time at least some newly emerged females also will use the site, until ever-larger aggregations of nests appear each year. Each nest is still worked independently, but for all outward appearances the area may appear to be one large colony of a bee or wasp species. Such groupings are called nesting aggregations, a purely artificial category, as it describes how adults of some bees and wasps appear to be living in a colony, when in fact they simply found an agreeable place to produce their offspring.

Once, when I worked at the Museum of Natural History in downtown Washington, D.C., there was quite a fuss when giant cicada killer wasps (Crabronidae) began appearing in huge numbers between our building and the neighboring Museum of American History. Cicada killers are among the largest—and to some viewers scariest—wasps commonly seen in the eastern United States, ranging up to 2 inches (5 cm) in length. As it turned out, they had been nesting unnoticed for some years in a grandiose bed of ground-covering ivy between our buildings. Eventually the numbers became so large as to be noticed by the general public, which, as could be imagined, was scared witless. As is usually the result in such cases of human terror, the entire colony was wiped out chemically to ensure the safety of the nation's capital. We'll return to cicada killers a bit later in this discussion.

In addition to large aggregations of a single species, sometimes a nesting area is suitable for a diversity of other species and groupings of several species of bees and/or predatory wasps occurs. On several occasions, I have seen sandy banks along small streams or ponds that were home to thousands of individual predatory wasps and bees of at least half a dozen different species. Sometimes dead tree trunks, wooden posts, old barns, adobe buildings, or clay cliff faces will be home to different species of boring bees or predatory wasps, all working independently, and each appearing unperturbed by the neighboring activity.

Colonial (or Social) Lives

In this category fall all the Hymenoptera that live together in a single nest that "consists of two or more adult females, irrespective of their social relationships" (Michener 2007). I hesitate to introduce more terms than necessary into this discussion, but I include here instances of intermediate behaviors that bolster the case for the evolution from solitary to social behavior in Hymenoptera. Not all authorities agree on the definition of these terms, so it is a bit tricky to define them, and I use the simplest definition possible.

In some bees and predatory wasps, females actually use the same entrance to a nest, for example, in soil or wood, with each building and provisioning her own cells. This is called communal nesting because each female is working independently within the nest but with no shared or divisions of labor. Put in human terms, each nest is equivalent to an apartment or condo, with females coming and going through a common entrance but each working in her own room (cell). Communal nesters provide no care of the young other than the original provision of food (prey or pollen), seal off the cell, and go on to build another cell in the same structure.

A step beyond this is quasisocial nesting, in which all females can lay eggs (as in the case of communal nesters), but females may perform "a particular task when she comes across a cell that requires a particular job" (O'Toole 1998). That is, there is cooperation between females: They do not just care for their own offspring, and any female can do any task.

In semisocial nesters, females of the same generation cooperate to build and provision a nest, there is an indistinct division of labor in which some females act as workers, and there is a principal egg layer, but there is no overlap of generations.

Because the status of communal, quasisocial, and semisocial behaviors are sometimes difficult to tell apart, the umbrella term *parasocial* has been given to cover these terms.

A few bees and predatory wasps are commonly referred to as social, as are all ants. In entomology the social bees, predatory wasps, and ants are given the special term *eusocial* because their behavior is entirely different from how humans define sociality. Basically (and there are exceptions) eusocial Hymenoptera have a single mother, the queen, who lays all the eggs. Her progeny are watched over and fed progressively by their own sisters. There is generational overlap so that new adult offspring (workers) assist their elder sisters in creating and maintaining a colonial formation. And, finally, there is a division of labor among working females.

One note of interest regarding eusocial insects has to do with how the queen rules her subjects. In the case of yellow jackets, hornets, and paper wasps, the queen rules purely by aggression. Worker females are bullied via head butting and physical charging by the queen, which may be referred to as "psychological dominance" (O'Toole 1998). Essentially it is a pecking order in which the queen reigns supreme. This aggression causes a worker's ovaries to remain undeveloped. A worker may occasionally escape the queen's wrath and lay eggs, which being unfertilized will result in male progeny. In the case of honey bees, bumble bees, and ants, the queen rules by subterfuge. She produces a chemical product, a pheromone, that acts as a drug spread to individuals throughout the colony. This pheromone controls the workers, turning them into the equivalent of mind-controlled zombies, not unlike the results of "soma" in Aldous Huxley's *Brave New World*.

The route from solitary to eusocial behavior is so full of exceptional and complicated steps that for predatory wasps alone the prominent wasp experts Howard Evans and Mary Jane West Eberhard (1970) recognized 10 different behavioral levels (plus nine additional subdivisions) and 28 different nesting types to account for the rise of eusocial behavior. The bees are scarcely different, except that they have achieved the highest level of any insect in the eusocial behavior of the honey bee.

As might be deduced from the many levels suggested by Evans and West Eberhard, categorizing the solitary and social behavior of Hymenoptera is quite an art, and it can prove difficult to make simple statements even about a single species of bee or wasp. Take the case of the cicada killer discussed above. Basically these are fine examples of solitary, predatory wasps. There are reports, however, that rarely two females actually use the same burrow but excavate separate cells within it (that is, effectively they act as communal nesters). A suggestion was also made (though not substantiated, apparently) that very small females, the sorts that would have trouble subduing a cicada, may actually sneak in and lay an egg on the contents of a cell being provisioned by a larger female (that is, they act as cleptoparasitoids; Holliday 2008). Within one species, then, it may be possible to find a few individuals that behave somewhat differently from the majority. Which is, after, all, one of the tenants of evolution: individual variation acted upon by natural selection.

In a few families it is possible to find several different sets of behavior among genera or species. For example, the predatory spider wasps (Pompilidae) attack their hosts and nest in several different ways. Some genera simply lay their egg on a free-living spider, others paralyze a spider in its own burrow, others dig a burrow, and yet others may have several females nesting together (communal nesters) and even overlap with their own offspring.

But enough for now, we shall discuss some of the specific and fascinating intricacies of the hymenopteran lifestyle in the second part of this book, wherein we examine the major groups of Hymenoptera in more detail.

4

A Garden of Hymenoptera

CURRENTLY IT IS a popular notion to garden naturally or to have wildlife gardens (see, for example, the excellent book *Bringing Nature Home* by Douglas Tallamy 2009). Generally, however, wildlife is merely a euphemism for two primary groups of animals: birds and butterflies. Thus, we have books on bird gardens or butterfly gardens, but seldom is mention made of lizard gardens, snake gardens, frog gardens, or even aphid gardens. The tendency is to accept birds or butterflies as being garden-worthy, whereas other creatures are simply ignored.

If one searches for "butterfly garden" on a web browser, the number of hits is over a million. Ignoring the notion that most of these sites are inane, useless, or pertain to some unfathomed aspect of pornography, a similar search on "bee garden" results in tens of thousands of hits, some of which are actually useful. But search for the item "wasp garden" or "parasite garden" and you'll barely find a couple hundred hits, most of which are have no bearing on the subject at all. "Sawfly gardens" are totally unheard of and for good reason: sawfly larvae eat plants (as do butterfly larvae, but scarcely anyone mentions that fact). Hits for "ant garden" don't really count because some ants build their own gardens, which is what the sites are about, not how to build a garden to attract ants. Based on this completely unscientific survey, we can see that butterflies in the garden are vastly preferred to bees, bees are somewhat preferred to wasps whether predatory or parasitic, and no one cares a whit about building a garden for ants or sawflies. It seems as if the usual state of human affairs is at work: an attraction for the bright, shiny aspects of nature in preference to the bugs that do the basic work of keeping our gardens functioning as nature meant them to.

I heartily agree with the desire to attract butterflies—I do it in my own garden—but in reality bees, predatory wasps, parasitoids, and even ants are much more beneficial for a garden's health than are butterflies. Even plant-eating sawflies serve a purpose. Although butterflies may be beneficial for a gardener's mental health, the truth is that butterfly larvae may actually eat some plants to the ground, leaving the garden in tatters. For the past two years, for example, almost every sunflower and its relative I have grown has been demolished by larvae of the bordered patch butterfly, which is a beautiful little sprite in spite of its feeding habits. Against all natural logic, relatively few of these larvae ate their normal, abundant native host plants consisting of several genera and species of sunflowers. Instead they consumed only the nonnative species and hybrids I introduced. My garden also housed dozens of delightful black swallowtails, which left behind hundreds of larvae that totally devastated patches of fennel, rue, and parsley. And now the artichokes are being eaten to the ground by painted lady butterfly larvae.

Butterfly gardens, it turns out, are not all they are cracked up to be, yet many books have been written advocating their existence. Conversely, wasps, ants, and even bees are generally greeted with the question, "How do I rid the garden of these threats?" I'm here to suggest instead that a better question might be, "How do I live with or even encourage bees, wasps, and ants into the garden?" Hymenoptera (as well as other insects) provide four basic services to a garden and the world we live in: food to attract critters such as birds, lizards, and frogs; balance to maintain their own ecosystem; pollination to produce seeds, fruits, and vegetables; and recycling to reuse organic wastes. This last service includes not only the recycling of nutrients (that is, movement of dead or living animal and plant tissues into the soil), but also the improvement of soil by aeration. These are services for which gardeners should have to pay huge sums of money, but they are provided free of charge if the gardener would encourage Hymenoptera or at least not discourage them.

As I noted in my previous book (Grissell 2001), when allowed to operate properly, these basic services help the garden function as a natural system and reduce problems such as the outbreak of unwanted pests. If our gardens were stable, with all creatures and plants in balance, we likely wouldn't know what a pest was because its numbers would be so low as to be completely overlooked. Gardeners don't realize that a "pest" often is simply one too many of an organism we hadn't even noticed before. The best way of improving a garden's stability is to increase its biological diversity (or biodiversity). Biodiversity is basically a measure of the number of different kinds of living organisms that inhabit a given space. The more diverse a given space becomes biologically, physically, and temporally, the more complex are the biological interactions between organisms. The more complex these biological interactions become, the more stable the entire system becomes. By increasing the biological stability of the garden, our lives as gardeners also become more stable, thus reducing our work loads and giving us more time to eat, drink, or make merry.

The basic methods of increasing hymenopteran diversity fall into two categories. One is an understanding of adult requirements, including attraction—provided largely by plants and moisture—and the avoidance of deterrents, such as pesticides. The other is an understanding of factors that induce the adults to take up residence and begin the process of producing offspring. Actually, these two factors are true for all insects, including butterflies. In the following discussions we'll examine ways in which the gardener can proactively increase the diversity of the garden with respect to bees, predatory wasps, and parasitoids. I've omitted ants and sawflies from this discussion for several reasons. Ants will enter the garden no matter what we do, so we needn't really bother to invite them. Immature sawflies are plant feeders, and although they might provide some nutrition for birds or frogs, they are more likely to be disagreeable to the gardener's basic instincts, so I've also omitted them. Besides, if you include certain common plants in your garden (for example, roses, pines, plums, cherries, peaches, willows, hibiscus, and purslanes), the sawflies likely will arrive anyway. Again, an invitation is not required.

For those who find living with Hymenoptera unpalatable, I will even—much against my better judgment—give some hints on how to exclude unwanted renegades of the order from your garden. After all, I occasionally attempt to be reasonable, even when it hurts.

Attracting Adult Bees, Predatory Wasps, and Parasitoids

Before any decent hymenopteran will take up residence in a garden, it is necessary for the adults to be drawn to it. Basically this first requires sustenance for the adult. Once adult requirements have been met, they then can be encouraged to stay and carry on with their reproductive activities.

Because adult bees, predatory wasps, and parasitoids are primarily nectivorous, they need nourishment in order to reproduce and develop eggs. So an initial attractant would be some sort of plant product, of which several are available. They also need a source of moisture, which is usually provided by nectar, but moisture can be supplied by the gardener with beneficial results. With few exceptions, almost any flowering plant (annual, perennial, shrub, vine, or tree) will attract some hymenopteran species when in bloom. But even when not in bloom, some plants have sugary rewards, called extrafloral nectaries, that attract bees and wasps. These may be found on leaf margins and petioles, leaf and flower buds, and even hairs along a plant's stems. Sometimes it is not the plant itself that attracts, but sap-sucking insects such as aphids or psyllids that excrete a sugary byproduct called honeydew. Predatory wasps and parasitoids are particularly attracted to honeydew, even when they have no interest in the insect that creates it.

It is customary when discussing butterfly gardens, for example, to provide lists of plants that prove attractive to adults—the glittery bits—though their larval needs are seldom taken into account. With the exception of bees, few plant lists are designed to attract adult Hymenoptera, and most of them are regional. Below I discuss some of these lists rather than produce a list of my own because I prefer to discuss general aspects of plant suitability that apply to Hymenoptera. There are some basic commonalities about plants that apply to these insects, or any insect for that matter, regardless of the region in which they grow.

In addition to the discussions that follow, I recommend two avenues of exploration that will prove useful. First, take these general comments about plants and look for regional guides to both native (such as field guides) and cultivated (garden books) species for the area in which you live. Second, visit your local nurseries and botanical gardens frequently and note the plants that are attracting the most hymenopterans, then use the same sorts of plants in your own garden.

Plant Lists

Whatever the eventual attraction, gardeners are well aware that plant species, hybrids, and cultivars can vary greatly by geographic region, and for this reason it is difficult to present a plant list within the confines of a book that would satisfy all gardeners. Many lists of butterfly- and pollinator-attracting plants already exist on the internet and in books that are geared toward general pollinators (such as beetles, flies, birds, bats, and bees), and these plants will also attract Hymenoptera.

Of the available pollinator lists that specifically treat Hymenoptera, most are slanted toward bees. Few compilers seem to consider it important to attract either parasitic or predatory wasps, and no one worries at all about attracting ants or sawflies. Fortunately, any plant that is attractive to bees is equally likely to attract predatory wasps. Parasitoids are largely attracted to plants with small flowers, and choosing such plants from the lists is useful in attracting this group of Hymenoptera. Many of the same plant genera and often species are found across these lists, which is not unexpected because garden centers, nurseries, and catalogs provide many of the same plants to the gardener's palette. Even in Arizona I now grow plants I once grew in Maryland and California. These plants attract Hymenoptera, which differ regionally in species complexes, but they still serve the same attractive purpose no matter the region in which they are grown. A plant listed for San Francisco may grow perfectly fine in New Jersey, but the plants will likely attract different sets of insect species.

After searching through dozens of potential pollinator lists, I've cited a few internet sites (see the list of useful websites) that relate to bees and plants that attract them. The most useful information for the United States is hosted by the Pollinator Partnership and their *Ecoregional Planting Guides*, which feature pollinator-friendly guides for 35 U.S. ecoregions, selectable by zip code. Other useful geographically broad guides are *Alternative Pollinators: Native Bees* and *Plants Attractive to Native Bees.* The Xerces Society, an organization dedicated to the conservation of invertebrates and their habitats, has produced nine guides for various regions of the country, the most inclusive being *Plants for Native Bees in North America.* In the United Kingdom similar lists may be found at the Bumblebee Conservation Trust, Bumblebee.org (lists are given for European plants in general), and the British Beekeepers' Association. In Australia the Aussie Bee website (hosted by the Australian Native Bee Research Centre) provides much general information about native bees, including a list of plants under their article entitled "Aussie Bee's Gardens Rise from the Ashes."

There is an excellent site for the San Francisco Bay Area (and similar Mediterranean climates) at Urban Bee Gardens. This site, produced by Dr. Gordon Frankie at the University of California at Berkeley, is well worth a visit as it explores the entire world of bees in an urban setting. It gives two seasonal lists of flowering annual and perennial plants for California garden as well as the bees that are likely to visit. For the mid-Atlantic region there is a site called "Pollinator-friendly Cut Flower Plants" (www.smallfarmsuccess.info/poll_friendly.cfm), which should prove useful for much of the eastern coast of the United States.

Floral Diversity

The subject of flowers is one dear to the heart of most gardeners. Many of us are in it for the flowers, but as is generally true of all things in our lives, they are but a small part of a larger whole. With regard to flowers, there is some controversy as to whether flower color, fragrance, or shape is what might first attract a passing insect. As with everything biological, there is likely to be some aspect of each factor or combinations of each that appeal to different insect visitors. For this reason alone, floral diversity is a requirement for designing a garden of hymenopterous delights. Let us explore flowers, then, as but a piece of a larger design in which our hymenopterans will be considered.

Size, shape, color, and fragrance

Generally small flowers attract almost any hymenopteran regardless of its size, including parasitoids, bees, and predatory wasps. Larger flowers, however, tend to attract large bees such as honey bees, bumble bees, carpenter bees and larger predatory wasps (not to mention butterflies and hummingbirds), but not many parasitoids. So in nature, size does matter. But it's the size of the individual flower not the appearance of the flowering cluster that is important. Sunflowers, for example, are composed of numerous small flowers, not one giant one, as in the case of lilies. I have seen sunflowers and their relatives (for example, *Tithonia*) with several different orders of insects on a single composite flower as well as three or four species of bees. In species of plants such as dill, parsley, anise, and fennel (Apiaceae) as well as catmints, oreganos, monardas, and mints (Lamiaceae), the individual flowers may be small but the clusters appear much larger. These latter two families are extremely attractive to Hymenoptera of all sorts.

Bees and wasps come in a huge array of different shapes and sizes. Some have long tongues, others have short ones. Thus, the more shapes of flowers found in a garden, the more and different kinds of hymenopterans will be attracted. Even within a single group such as bumble bees, there are long-tongued species that can reach directly into the nectar spur of a tubular flower such as a columbine, whereas short-tongued species will simply bite a hole at the base of the tube to reach the nectar. Most bees apparently prefer bilaterally symmetrical flowers, by which they orient themselves and then enter the flower's throat. Shallow, dish-shaped flowers, such as apple or blackberry, which are radially symmetrical, are called open-access flowers because many different sorts of insects, including Hymenoptera, have access to the nectar.

Bees and wasps see color differently than humans. Their world is made up largely of whites, yellows, blues, and greens, but true red appears as black. They have the ability to perceive the ultraviolet range of light unseen by humans, so if there is an ultraviolet component in a red flower, it will still be a useful inducement to them. Some flowers, although they appear entirely yellow or white to us, have patterns that reflect ultraviolet light as a bull's-eye centered around the reproductive area or ultraviolet landing lines that act as guides to the flower's center. These guideposts are for visiting pollinators, and we humans are none the wiser for it. Basically, then, almost any color will attract bees and wasps, though pure reds would be the least attractive unless they have an ultraviolet component.

The primary function of scent in flowers is to attract pollinators, although scent may also play a role in defending them from plant feeders as well. A flower that secretes a scent, even if we humans cannot detect it, is more attractive to bees and wasps than a scentless one. According to a recent survey, nearly 1800 different floral chemical compounds were reported from nearly 1000 different plant species, and the authors contend this is but a small sample of the floral kingdom (Knudsen et al. 2006). Obviously we humans are poorly informed when it comes to floral scents, but as we shall see below when we discuss patches, chemicals in general play a large role in the lives of insects.

Floral scents can emanate from glands situated on petals or from pollen and nectar itself. Surprisingly, some flowers do not provide nectar inducements to their visitors and thus must rely on some other bribe, such as fragrance. Because insects have the ability to orient to chemical signals of which we are totally unaware, it is difficult to know how to select flowering plants based on fragrance. In such cases, I simply recommend selecting plants that appeal to our own sense of smell as well as adopting the principle of diversity. If you have floral diversity, then something is bound to work whether we plan for it or not.

Single, double, and pollenless flowers

Single flowers are generally preferred food sources for Hymenoptera. In single flowers the nectaries, stamens, and anthers (pollen-bearing organs) are easily accessible and are present in their most numerous configuration. This allows Hymenoptera full access to nectar and pollen. In double or multipetalled flowers, however, the stamens have been transformed into petals (thus no pollen) and the nectaries, if present, may be totally buried in a mop-head of petals. At some point, for example, in extremely double-flowered zinnias or dahlias, none but the larger bumble bees or carpenter bees can reach the nectaries, providing they can sense they are present.

In addition to multipetalled flowers in which the stamens are transformed to petals and the pollen has been reduced, some plant strains have been bred specifically to eliminate pollen (some cut flowers such as sunflowers). These plants appear normal, but are missing half their attractive potential. If nectaries are present, adult Hymenoptera might be found supping for themselves, but no self-respecting bee would be caught working such flowers as there would be no pollen to take back to provision its nest.

Seasonality (floral phenology)

Although the gardener's impression may be that honey bees or yellow jackets, for example, are foraging from spring to autumn, this is actually rather unusual behavior for Hymenoptera. The solitary sorts of bees, predatory wasps, and parasitoids generally have limited, season-defined life cycles during which they can reproduce. Within Hymenoptera, populations of different species vary throughout the year so that certain bee species, for example, are active only in spring, some in midsummer, and some in late fall (Figures 32–37). Therefore it makes sense to create a garden that has some bloom during the entire growing season. Most gardeners do this anyway to soothe their own psyches, but floral displays that vary in time also create opportunities for a larger variety of Hymenoptera to go about their own survival. Just as adding a diversity of flowers encourages a variety of bees or wasps to visit, so, too, does a continuously varying supply of flowers over the longest possible time aide the largest number of bee and wasp species in taking advantage of a garden.

Plant Diversity

Flowers are but one aspect of gardening for bees and wasps. Equally important is the selection of plants that produce these flowers as well as their manner of placement. After all, a garden is not merely the sum of its flowers. Whether grown as a collection of botanical subjects (that is, a collector's garden) or created as a designer's abstraction, it is a living entity subject to the rules of nature and certain ecological principles.

Plant diversity plays an important role in how a garden relates to the natural world around it. A garden of roses or hostas might be termed a collection of plants. It may appear to be diverse based on flower color and plant architecture, but such a collection really can't be considered biologically diverse. An insect that attacks roses, for example, will find its heaven in a rosarium. Endless hostas will produce the finest slugarium imaginable. Neither garden will attract a broad spectrum of insects by itself, but will instead act as attractant to those few and particular insects that enjoy its overwhelming presence.

Creating a garden that enhances the likelihood of attracting adult bees, predatory wasps, and parasitic wasps merely requires that it be a garden of diversity. To achieve diversity, a garden must have many different kinds of plants including native and cultivated species; flowers of different sizes, petal count, shapes, colors, and seasonality; and integration of plants into a network that attracts so much biological diversity that attacks on its structure are absorbed with scarcely a notice by the gardener. As an example of the consequences of a lack of diversity I note my experiences with a plan to attract butterflies and Hymenoptera using a single species of plant. I planted a patch of fennel, some 15 by 6 feet (4.5 by 1.8 m), as an inducement for black swallowtails to lay eggs. The patch attracted a dozen or so adult swallowtails that laid so many eggs the plants rapidly turned into bare stems that never had a chance to flower and attract Hymenoptera. Let this be a lesson, then, that sometimes it's difficult to attract only the insects you want without inviting some of those you might not want. Making biology work exactly the way you want it to is not easy, no matter how much you might plan.

Patches

As we have just seen, the concept of patches introduces a major Catch-22 into a gardener's life. In ecologist's jargon a *patch* is merely a grouping of plants of the same type. Patches in gardener-speak are swaths, beds, borders, or plots, and are commonly suggested as the appropriate way in which to stage plants in the garden. They are often called for in esthetic landscape designs where plants are clustered to provide impact and to awe the viewer. Patches are also the hallmark of vegetable gardens, where convenience of cultivation is important. It is tempting as a gardener to plant banks of luscious roses, swaths of salvia, beds of zinnias, or rows of squash. To our eye these clusters are alluring, orderly, and satisfying. Assuredly they will attract bees and wasps of many kinds, but there is a catch involved with clustering. That catch is biology.

Plants, as it turns out, are not the least bit impressed with human design because they live in a world dictated by biological function. Without regard for human plans, plants diffuse chemical signals from their leaves and flowers that may cause insects to orient to them. When plants are set out in patches, that chemical signal is multiplied and the patch becomes more attractive to insects. Now this is a good thing if you want to attract bees to pollinate them or predatory and parasitic wasps to restore balance to a garden. It's not so good, however, if what you attract are insects that want to feed directly on the plants in the patches.

Once sensing an initial chemical signpost, some leaf-feeding true bugs will fly toward the signal, landing first on one plant, then another, until they have touched down on at least three or four different plants of the same type before they begin feeding for themselves and laying eggs. Thus, a patch of a single kind of plant assures an adequate supply of food for both adults and young. In many cases, however, insects are following an odor trail not for their own benefit but because it means dinner for their larvae. Butterfly and moth larvae fall into this category. A female cabbage butterfly, for instance, orients to the fragrance of cabbage foliage although she, herself, does not feed on cabbages. She is looking for a place to lay eggs. The more cabbages in a patch, the more chemical signals are produced, the likelier a cabbage butterfly will find it, and the likelier she will lay eggs.

If the same plant can attract both desired and undesired visitors, and this effect is magnified by planting in patches, how is a gardener to overcome the conflict? One commonly advocated plan to surmount the attractiveness of patches is to place companion plants around or within (that is, interplanting) them. One of the main tenants of this philosophy is that such plants will repel unwanted insects, usually referred to as "pests." Another possibility is that the companions will become sacrificial lambs that are so attractive they will entice invading insects away from the protected plant and onto themselves (referred to as trap or catch crops). Although both scenarios are possible, the most likely aspect of companion plants is that they act as refugia wherein parasitic and predatory wasps find protection and food and from which these beneficial insects then will emerge to protect the subject patch from invading insects. In point of fact, companion planting is simply a way to increase plant diversity.

Native plant species

It certainly makes sense to introduce native plant species to a garden because many local Hymenoptera (especially bees) are specifically adapted to seeking nectar and pollen from flowers of these plants. As native plants are all the rage these days, acquiring them is becoming less difficult than in days of yore. I've found locally grown native plants as close as my hardware store, though these natives may have been propagated in Switzerland for all I know. Just what is a "native plant" is open to much speculation, and I suggest you find some other book to help you with that dilemma. I'm having difficulty with the concept myself even though I belong to two native plant societies.

Water Sources

Most adult bees and wasps require an uptake of moisture for survival. The source can be as simple as dew on a leaf, a mud puddle, or a pond's surface. Therefore, providing a water source in the garden will be attractive to many bees, wasps, and insects in general. Devices such as shallow saucers of water, birdbaths filled with moist sand, or even muddy areas will be attractive to some form or other of Hymenoptera. Some wasps will take up water directly and then mix it with soil to form mud nests or with plant fibers to form paper nests. Others will take up mud directly to form their nests. So free water as well as muddy puddles will serve as attractants to hymenopterans. If you will permit a personal note, I'll digress on the issue just a bit further.

Gardeners are well known for stealing ideas (and cuttings, as well) from the gardens they visit, and I'm no exception. While visiting the Oklahoma State University Botanical Garden in Stillwater, I was given a tour by Kim Rebec, assistant extension specialist and host of the Oklahoma Gardening Program. Nestled in one corner of the garden was a small, in-ground fountain that had been set up as a water source to attract butterflies. This was no gushing monstrosity, but simply a tiny pump sending a bit of water an inch or so into the air—just enough to moisten the surrounding soil a bit. Upon returning home I duplicated the fountain with modification, ostensibly to attract butterflies (Figure 38). Instead of using an electric pump, I found a small, solar-powered pump online. After digging a pit in which to sink a water-holding container at ground level, I placed two lengths of wire shelving across the top with a space in the middle so I could stick my hand into the container to service the pump if needed. The solar pump went into the container, which was then filled with water. Rocks were arranged over the shelving to hide everything and to provide a surface upon which the water could splash as it fell back down into the hidden container. Interestingly, as the sun moves across the collector, the water height varies from a trickle to a splashing display at midday, then back to a trickle again in late afternoon. Thus a variety of rock wetting occurs, which seems to appeal to differing insect crowds (Figures 39–44). Some wasps seem to enjoy getting right into the splatter of things, to the point where they have to dry off in order to fly away, whereas the butterflies are more content when there is more moisture than splatter.

As I say, this setup was ostensibly created to attract butterflies, which are well-known mud-puddling creatures. A few different species did occasionally settle upon the rocks, but to my amazement the water-coated rocks were more attractive to a variety of bees, predatory and parasitoid wasps, and ants than I had ever imagined. Coincidentally, many of the wasps are predators or parasitoids of caterpillars, which should be of some interest to those wishing to improve natural control in their gardens.

Pesticides

Having planted a fabulous garden devoted to natural diversity, the one final aspect of attracting adult bees and wasps is to not make their lives miserable once they get there. Here the biggest mistake that can be made is the use of pesticides. The broad-scale use of synthetic pesticides to control insects in general will certainly reduce populations of most Hymenoptera. It does so either by poisoning them directly or indirectly by killing off the prey and hosts they need for their progeny to survive. Hymenoptera are especially susceptible to chemicals because of their small size (doubly so in the case of parasitoids) and high rates of metabolism. Organic insecticides such as the various bacterial or fungal derivatives (for example, *Bacillus thuringiensis*) will affect Hymenoptera, again because they kill off the hosts.

Even the benign, nonchemical control of one species of insect can be detrimental to the health of nontarget species, and I present a cautionary tale to illustrate my point. I have a greenhouse attached to my house here in Arizona in which I began experiencing problems with whiteflies, mealybugs, aphids, and fungus gnats. My first inclination was to go the mechanical route and handpick the offending invaders. This did little to solve the problem, so I set out several yellow sticky traps, especially for the whiteflies. To my horror, the very first insects I found on the traps were parasitoids of the whiteflies. These are barely visible wasps about 1/16 inch (1.5 mm) in length. Then I found a few whiteflies on the traps, so I felt a bit better. But then to my even greater horror I discovered dozens of aphid parasitoids stuck to the trap. Upon closer inspection of the living aphids, there were hundreds of parasitoids going about their business of aphid slaughter, and I opted to remove the sticky traps. Although the aphids finally disappeared due simply to the slaughter, I was still stuck with whiteflies, mealybugs, and fungus gnats, which I eventually controlled by dumping all the infested plants in the compost pile. The lesson here is to give nature a chance (it worked with the aphids and has done so before in similar situations), but when the battle is lost admit defeat and move on with your life.

Habitat Requirements for Hymenopteran Progeny

In order to successfully reproduce, Hymenoptera must have suitable environments in which their progeny can develop. As with butterflies, merely attracting passing adults does not ensure survival of the species. There must be an inducement to stay and procreate. In the case of bees and predatory wasps, this requires an appropriate nesting site (as discussed below) where progeny are encased and protected as they grow. A carpenter bee, for example, nests in holes it excavates in wood, whereas a cicada killing wasp nests in burrows it excavates in the ground. In the case of parasitoids, which don't build nests, the female must find the exact type of host at the correct stage in order to lay eggs and allow its progeny to develop.

Obviously a gardener cannot expect the garden to replicate every requirement for every bee, predatory wasp, or parasitoid in the region. A common-sense approach will work just fine, which is why the subject of garden diversity again rears its inquisitive head. A garden of both plant and habitat diversity will ensure that some hymenopteran will find its way into the mix. To illustrate this point I turn to my friend and colleague Dr. Frank Parker (a former research entomologist at the USDA Bee Biology and Systematics Laboratory), who has extensively studied solitary bees and wasps. One of the techniques he uses is to place stick traps and block nests in both natural and garden situations (Figure 45). In one of his research projects he sent block nests to 300 collaborators, who placed them in their gardens throughout the United States and Canada. Of these, 270 blocks were returned from 46 states and two Canadian provinces. Included in the blocks were the nests of 2006 bees and wasps.

From these nests Dr. Parker recovered nine genera of bees representing about 20 species, with both flower specialists and generalists present. Many of the species were primarily pollinators of composites, a few were legume pollinators, but most pollinated many different species of plants. The bees, however, represented slightly less than one-third of all the nests studied. The most surprising result of his study was that predatory wasps, not bees, were the primary residents of the block nests. Although these wasps were too diverse to identify to species, some rough idea of their impact in the garden can be gained by looking at the host insects they feed upon.

One-third of all the nesting Hymenoptera found in the block nests were crabronid wasps (Crabronidae), which use many different insects as prey for their offspring, including immature grasshoppers, aphids, planthoppers, and spiders. The second largest group of wasps, equal in numbers to the bees, was the eumenids (Vespidae), which prey exclusively upon leaf-feeding insects. Most commonly they use moth larvae, but some species specialize in leaf-mining beetle larvae. Rounding out a few odd percentage points were predatory sphecid wasps (Sphecidae), which prey upon crickets and spiders, including the black-widow. And finally there were true spider wasps (Pompilidae), which also use spiders as hosts for their progeny.

It is obvious, based on Dr. Parker's relatively small sample of garden nesting blocks, that much unseen activity takes place whether the gardener plans it or not. Thus, the endeavor of creating a garden of hymenopterous delights is perhaps not as complicated as it might seem because successful activity is already taking place about which the gardener is completely oblivious. The solution to meeting the reproduction requirements of many Hymenoptera, then, is simply to have a garden of diverse habitats.

Parasitoids and Their Habitat Requirements

The case for inviting parasitoids into the garden is relatively simple compared to keeping them there. Whereas adults may be found foraging at suitable flowers for their own nutrients, what really drives them is the search for hosts they can parasitize. But their hosts generally are not what gardeners want in the garden. If you want aphid parasitoids, for example, your garden will require aphids, or at least some aphids nearby, as in your neighbor's garden. This is where the very delicate subject of natural balance becomes involved.

In order for balance to work in favor of parasitoids, there must be at least some of their hosts available. Parasitoids are generally extremely catholic in their tastes, and a cabbage worm parasitoid, for example, will not attack an aphid. The ideal situation is to have a low level of host population constantly in the garden, which in turn harbors a low level of parasitoids. This insures that if the host population begins to build up, then the parasitoid's population can easily keep pace with it, maintaining everything in balance. When the gardener panics at the first sign of a problem, he or she is likely to be compounding it by attempting to eliminate the host. Thus, gardeners, who are some of the most patient people I know, will do well to practice extreme patience when it comes to some insect infestations.

Predatory Wasps, Bees, and Their Habitat Requirements

When speaking of predatory wasps and bees, we basically are discussing the female of the species, as it is she who is indefatigable when it comes to finding the exact habitat in which to rear her young. Bees and wasps as a group inhabit almost any environment, natural or manmade, and thus may appear to be indiscriminant in their nesting requirements. This, however, is a misleading picture, as individual species are highly specialized when it comes to using specific nesting spots in which to build and provision nests for their young. The comments here are generally for solitary wasps and bees, although they also apply to the few social sorts that will commonly be found around the garden.

In general, species of both predatory wasps and bees require habitats that fall into three basic categories: species that nest in the soil, species that nest in wood or plant parts, and species that build their own structure (for example, a mud tube) to house their young. This section deals with creating living conditions for predatory wasps and bees, so I needn't go into details of specific nest architecture—those discussions will appear in the second part of the book. Instead, we'll examine the types of habitats that would prove inviting to our hymenopteran allies.

Ground nesters

Many thousands of solitary wasps and bees nest in the soil or ground. As gardeners you might realize that soil is a highly disparate commodity, varying in texture, chemical composition, and even dimension. With regard to dimension, some species of ground-nesting bees and wasps nest only in vertical soil banks, such as along the walls of a canyon or stream. Others nest in slightly banked or sloped soil, and yet others nest only in flat areas. Compounding the factor of dimension is that some species of bees or wasps only nest in level sandy soil, whereas others prefer compacted, level clay soil. This even applies to vertical banks: Some can be quite unstable sandy cliffs and others rock solid and seemingly impermeable. An added complication to nesting requirements is that some species of bees and wasps will only use sandy level soil if it is totally bare and open to the sun, other species require the presence of a surface cover of organic debris such as leaf litter, and yet others require low-growing, sparsely placed plants to satisfy their requirements.

As some practical examples of ground-nesting sites used by bees and wasps, one of the smallest, yet most diversely populated sites I've ever encountered was along a barren, firmly packed, sandy stream bank. Here, hundreds of bees and wasps of many different species were all using a space no larger than 30 by 10 feet (9.1 by 3 m). It seemed impossible that all these solitary female nesters could avoid either banging into each other as they flew madly about, or that they could even find their own nests because there were so many holes in the sand. Generally, nesting aggregations such as this are confined to a single species of bee or wasp that have built up over a series of years. For example, a huge aggregation of cicada killing wasps was found near my workplace in downtown Washington, D.C. These wasps were nesting in clay soil nearly covered by English ivy vines. I've seen other examples of sparse suburban lawns in which hundreds of individuals of a solitary bee species were nesting. The point is that almost any soil or orientation will fit the needs of some nesting wasp or bee.

My advice to gardeners who want to encourage ground-nesting bees and wasps is simply to allow some soil diversity in your garden. Small, bare patches of ground (either clay or sand) are not necessarily a bad thing as they may be used by nesting bees or predatory wasps, which provide free pollination or insect control. If bare ground is too disconcerting, cover it loosely with rough-chopped herbaceous mulch and let a bit of soil show through. The trick is not to disturb the soil, especially if it is being used by nesters, because they will come back year after year if the site is relatively undisturbed.

Wood nesters

Many thousands of species of predatory wasps and bees are also known to nest within plant parts such as stems, dead twigs, and tree trunks. In the case of plant stems, many of these wood nesters are capable of chewing into the soft, inner cores provided by some woody stems, such as rose canes, or accessing the hollow inner cores provided by others such as bamboo. Generally these stems will be dead and in a dried state, having been cut or broken off. It is difficult to chew into a living stem with its potentially clogging sap flow, so old, dried stems are the choice of those bees and wasps in the know.

In the case of really woody materials, such as tree branches or dead trunks, these nesters reuse the burrows created by other wood-boring insects, such as beetles, which have exited the wood. In a few cases, such as carpenter bees, the female is able to chew her way into almost any sort of exposed wood including structural members. In reality, these wood nesters are simply using a substrate that provides space in which to drill a cavity of some sort or to usurp a cavity created by another insect. In this respect, they are not to be feared by gardeners as plant destroyers.

As a result of the tendency to nest in hollow places, enterprising gardeners set out bundles of soda straws, bamboo canes, or blocks of wood with holes bored in them to act as nesting receptacles for bees and wasps. Garden catalogs are now selling such devices if you don't want to make them yourselves. As with the variety of nesting sites for ground nesters, nesting devices will attract different wood-nesting subjects depending on their placement and the entrance size. Some bees, for example, prefer the holes to have horizontal orientations (that is, cliff nesters), whereas others prefer vertical orientations (that is, ground nesters). Differing sized bees and wasps prefer openings of various sizes. Thus, as with all things biological, diversity again rears its bewildering head, but things need not be overly complicated. A few years back I discovered a tiny predatory wasp burrowing into the dried flower stalk of some parsley I hadn't yet cut back. The stalk contained several completed cells, each with a wasp larva set amid a cluster of tiny adult flies. There were only a few such parsley stalks, so a garden need not be a monument to dead stalks in order to attract predatory wasps or bees. It simply needs to be hospitable, by which I mean a bit messy, perhaps, or at least not entirely scraped bare of every bit of apparently useless organic matter.

Builders

Builders construct their entire nest from raw materials, which they gather or manufacture. One common example of a builder is the mud dauber (Sphecidae), a wasp that collects mud and creates elongate mud pods in which to store spiders and lay its egg. Potter wasps (Vespidae) also use mud but are much more refined in their designs. They construct perfectly proportioned mud vessels, which the English naturalist John Crompton (1955) described as "vases of earthenware that the Greeks might have envied." In these, the potter wasp places her caterpillar prey and lays an egg. Some social wasps (Vespidae) such as hornets and paper wasps collect bits of plant fiber, chew it, and make papier-mâché nests, either entirely open to the elements (for example, paper wasps) or enclosed in an aerial paper case (hornets). Some bees (for example, resin bees, Megachilidae) create nursery cells stuck to rocks by plastering plant resins with tiny pebbles to create a nearly rock-hard case. Thus, all sorts of raw materials from mud and pebbles to plant fiber and plant resins serve to benefit some form of predatory wasp or solitary bee. None of this usage is detrimental to the garden, its plants, or its gardener.

Opportunists

As with all things biological, there are some creatures that defy easy classification or fall into several different groupings even within the same species. They don't easily fit the three categories of ground nesters, plant nesters, or builders and simply do things their own damn way, which makes for very sloppy bookkeeping, if you know what I mean. Basically opportunists are somewhat like remodelers, taking advantage of preexisting structures that they modify to suit their own specific needs.

Honey bees, for example, are probably best considered opportunists because they generally need some sort of structure in which to build their nest. This can be a hollow tree, an attic, or even a purpose-built bee hive. In my Maryland garden I even saw honey bees build an open set of combs hanging from my apple tree. Once safely inside a cavity, the actual nesting space is structured from secreted wax pellets that are formed into honeycomb. Thus, honey bees are a sort of wood-nesting builder.

Bumble bees are generally considered ground nesters, preferring to use abandoned rodent burrows, but I've seen them build nests in a bird house and I've heard tell of them nesting in abandoned mattresses and armchairs. Some suppliers even sell artificial bumble bee boxes designed to induce them to nest. (I've not had any success with such boxes, but I'm told they can work.) Once a bumble bee finds a suitable habitat, it will build a nest composed of globular wax cells. Bumble bees might be termed ground-nesting builders.

Unlike their fellow social vespids (that is, hornets, paper wasps), most species of yellow jackets seek out or excavate hollowed, underground cavities in which to build paper nests. In my former garden, however, I found a nest suspended just above ground level, which was about as unusual as the free-hanging honey bee nest just mentioned. (Actually, the nest found me as I was mowing the lawn and received a few wasp stings for my effort.) I've seen photos of huge yellow-jacket nests built on the sides of trees and inside of abandoned automobiles. Thus, yellow jackets are ground-nesting opportunists, excavators, cavity fillers, or occasionally nest hangers.

Figure 1. Rarely examined closely, wasps are complex, beautiful, and interesting creatures. This parasitic ichneumonid wasp (Ichneumonidae), pausing to clean her antennae, is an example of the largely ignored world found in our own gardens. Photo by Carll Goodpasture

Figure 2. Hymenoptera are often difficult to tell apart, especially on the wing. Sometimes knowing the host of a wasp will help place it to family. Here are two examples of solitary predatory wasps in different families: a spider wasp (Pompilidae; left) that provisions its nest with spiders and a sphecid wasp (*Podalonia*, Sphecidae) that provisions its nest with cutworm larvae. Sphecid photo by Carll Goodpasture

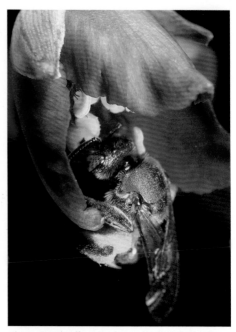

Figure 3. Unfortunately, yellow jackets (*Vespula*, Vespidae) are extremely well known examples of Hymenoptera, though they are vastly outnumbered by other species of the order. Yellow jackets account for only about a dozen of the 22,000 species of North American Hymenoptera. Photo by Carll Goodpasture

Figure 4. Small green bees (Halictidae) of many species may be just as common as honey bees, but they are less well known. Photo by Carll Goodpasture

Figure 5. Face to face with a true killer—a cicada killer, that is (*Sphecius*, Sphecidae). These predatory wasps hunt cicadas to place in their underground nests as provisions for their immature stages. Photo by Carll Goodpasture

Figure 6. If any good press is given to bees, wasps, and ants, it almost certainly goes first to the honey bee (Apidae), one of the most studied invertebrate animals in the world. Honey bees have been color coded, bar coded, numbered, and even radio collared in an effort to understand and manipulate them better. In this instance, the queen has been marked with a blue spot. Photo by Carll Goodpasture

Figure 7. When it comes to pollinating crops, honey bees (Apidae) often receive the glory when they aren't even present. This solitary andrenid bee (Andrenidae) is pollinating an apple blossom, though most would assume it's a honey bee. Photo by Carll Goodpasture

Figure 8. Not all pollinators are bees, though this cimbicid sawfly (Cimbicidae) appears as if it might be some sort of pollinating bumble bee. Photo by Carll Goodpasture

Figure 9. When they're not busy cleaning themselves, predatory wasps collect insects to feed their young. This paper wasp (*Polistes*, Vespidae) is a social species that carries masticated prey, often caterpillars, back to its paper nest and regurgitates the food, much as does a bird.

Figure 10. Even if we rarely see them, parasitic wasps play the most important role in controlling other insects and account for the greatest number of hymenoptera species. This ichneumonid wasp (*Xorides*, Ichneumonidae) is a parasitoid of wood-boring beetle or sawfly larvae. A long ovipositor is an indication that such wasps search for hosts in deeply concealed positions. Photo by Bob Carlson

Figure 11. Ants (Formicidae) offer up a great amount of biomass for the consumption of other animals, both vertebrates and invertebrates. Photo by Carll Goodpasture

Figure 12. A dahlia flower provides both gardener and bumble bee (*Bombus*, Apidae) with inspiration. Photo by Carll Goodpasture

Figure 13. This sweat bee (*Halictus*, Halictidae) has an impressively colorful gaillardia landing pad. Photo by Bob Carlson

Figure 14. Recently a eucalyptus boring beetle was accidentally introduced into California and is causing mortality and destruction of eucalyptus trees. This Australian wasp (*Syngaster lepidus*, Braconidae) is being tested to see if it can control the beetle larvae while not harming any native California beetle species. This is an example of parasitoids as potential biological control agents. Photo by Dong-Hwan Choe, University of California, Bugwood.org

Figure 15. Members of the suborder Symphyta, as exemplified by this hibiscus sawfly (*Atomacera decepta*, Argidae), are composed of a head and body with no constriction between the thorax and abdomen. There is no propodeum or petiole as compared to the suborder Apocrita. Most immature sawflies appear much like caterpillars and are plant feeders. Length 1/4 inch (6 mm).

Figure 17. In this extreme example of a thread-waisted wasp, the propodeum is just above the hind legs but is indistinct because it is the same color as the thorax. The petiole (second abdominal segment) on this wasp is extremely long and thus easily seen. This is a predatory species (*Ammophila*, Sphecidae) that hunts caterpillars for its young. Length 1 1/4 inch (32 mm).

Figure 16. Members of the suborder Apocrita, as exemplified by this gorytine wasp (Crabronidae), show an apparently distinct subdivision between the abdomen and thorax. However, the rounded, orange-and-yellow hump above the hind legs is the true first abdominal segment, the propodeum, which is fused to the posterior of the black thorax. This peculiar construction separates all parasitic wasps, predatory wasps, bees, and ants from all other insects. The articulating petiole (second abdominal segment) cannot be seen in this wasp because it is so small. Gorytines are predatory wasps that hunt leafhoppers and their relatives. Length 1/2 inch (12 mm).

Figure 18. In sawfly adults (*Ametastegia articulata*, Tenthredinidae) there appear to be only two body divisions, the head and the remainder of the body. There is no subdivision of the thorax and abdomen because they are fused. Length 1/4 inch (6 mm).

Figure 19. Because parasitic wasps come in many sizes, shapes, and designs, it is impossible to show a representative wasp. This is a common but bizarre and seldom seen eupelmid wasp (*Metapelma*, Eupelmidae) that parasitizes larvae of wood-boring beetles. Length 1/4 inch (6 mm). Photo by Michael Gates

Figure 20. *Chlorion* species (Sphecidae) are cricket hunters. Some species collect and place crickets in an underground cell. Others dig down into a cricket burrow, either paralyzing it in place or chasing after it, should the cricket escape. The wasp lays an egg on the host, which sometimes recovers from paralysis and consigns itself to its burrow never to leave again. Length 1 1/4 inch (32 mm).

Figure 21. Whereas wasps (*Tachytes*, Crabronidae; left) are relatively hairless, bees (*Bombus*, Apidae) generally have a fuzzy appearance. Additionally, wasp hairs are unbranched, whereas bee hairs are branched. However, there are examples of both wasps and bees that are essentially hairless. Photos by Carll Goodpasture

Figure 22. Megachilid bees (Megachilidae) are specially constructed to carry pollen on the underside of their abdomen, not their legs as is the case in most bees. Like most bees, however, megachilids are solitary, building individual cells in which they deposit pollen, lay an egg, and then seal and abandon the cell. Length ³/₈ inch (9 mm).

Figure 23. Ants (Formicidae) are likely the most abundant insects on Earth and easily among the most recognized of all hymenopterans. Several species of harvester ants (*Pogonomyrmex*) are the most common supplied with ant farms. Length ³/₈ inch (9 mm).

Figure 24. Female trigonalid wasps (*Taeniogonalos*, Trigonalidae) lay eggs that must be ingested by a plant-feeding caterpillar or sawfly larva. The wasp larva will then only feed within the host larva if it has been parastized by some other insect. Length ³/₈ inch (9 mm).

Figure 25. A female (left) and male (right) of the same species of velvet ant (*Pseudomethoca frigida*, Mutillidae), showing the extremes of sexual dimorphism. Such variants make identification of adult Hymenoptera extremely challenging, even to the family level. Length ¼ inch (6 mm).

Figure 26. Hidden from view, uncounted lives are being lived in this Arizona oak woodland. In the oaks are various species of gall wasps (Cynipidae) and their parasites. The junipers are infested with gall-forming flies (Cecidomyiidae) and their wasp parasitoids. Skunkbush (*Rhus trilobata*) and pines have seed-feeding wasps. A dead tree harbors beetle larvae and their hymenopteran parasites. Solitary bees and predacious wasps may be nesting in abandoned beetle burrows or between crevices in rocks or in the ground.

Figure 27. Galls, caused by plant-feeding gall wasps (Cynipidae), come in many shapes and sizes. Some are both intricate and colorful, as are these examples caused by *Andricus tecturnarum* on *Quercus arizonica*.

Figure 28. Seeds of many plants are infested by seed-feeding wasps belonging to the superfamily Chalcidoidea, primarily known for its huge numbers of parasitic species. Sumacs are a common host, with skunkbush or squaw bush (*Rhus trilobata*) being one of the favorites.

Figure 29. Solitary predatory wasps nest in many different ways. This sequence shows the nesting procedure used by *Ammophila procera* (Sphecidae) nesting in the sand between flagstones. Ammophila are the only wasps known to use an implement (that is, rock) as part of their nesting routine. The wasp begins excavating her underground burrow using her jaws and legs. Before the wasp leaves to hunt prey, she first closes the nest with a rock, then scatters sand over its surface. The wasp returns with a single moth larva, which she has dragged over the ground. The wasp reopens the nest, temporarily removing the rock. She turns around, grabs the larva, and backs down the burrow, dragging it to a cell belowground. The wasp lays an egg, then exits the nest, whereupon she replaces the rock, smoothes out the soil surface, and departs—her job finished.

Figure 30. Two ichneumonid wasp (Ichneumonidae) cocoons are seen here. This wasp is an endoparasitoid because its larva lives inside the caterpillar host. In this example, the parasitoid larva has exited from its host, then spun an external cocoon in which to pupate. Photo by Carll Goodpasture

Figure 31. The emerging braconid wasp (Braconidae) shown here fed as a larva inside the moth caterpillar. Once mature, this endoparasitoid pupated within the host, then chewed its way to the outside, leaving nothing more than a shell of its former host. On the left, a ragged hole indicates that another wasp has emerged from the other larvae. Photo by Carll Goodpasture

Figure 32. By late autumn, this Mexican sunflower (*Tithonia fruticosa*) appears the worse for wear as all its ray flowers have been devoured by grasshoppers and the occasional katydid. Still, the disc flowers are loaded with nectar and serve as a simultaneous landing platform for a diversity of insects, as shown here and in Figures 33–37. This flower accommodated a green bee (Halictidae), a wood-boring beetle, a checkered-skipper, and a honey bee (Apidae).

Figure 33. Three male *Halictus* bees (Halictidae), a honey bee (Apidae), and a grasshopper all find sustenance.

Figure 34. An anthophorid bee (*Anthophora*, Apidae) goes face to face with a checkered skipper butterfly.

Figure 35. As a grasshopper and katydid continue to decimate the ray flowers, a small green bee (Halictidae) still manages to find nectar.

Figure 36. A checkered-skipper and megachilid bee (Megachilidae) have no problem finding nectar on this flower with one remaining petal.

Figure 37. A large paper wasp (*Polistes*, Vespidae) and small male green bee (*Agapostemon*, Halictidae) coexist in their search for nectar.

Figure 38. A source of water encourages insect visitors. This small solar fountain provided moisture for the hymenopterous visitors shown in Figures 39–44. By coincidence the wasps illustrated here all use lepidopterous caterpillars or pupae as food for the development of their young.

Figure 39. A species of vespid wasp (Vespidae) commonly referred to as a euminid. These wasps nest in preexisting cavities.

Figure 40. Eumenid wasps (Vespidae) come in a variety of sizes, colors, and patterns.

Figure 41. A eumenid wasp, *Eumenes* (Vespidae), called a potter wasp because it uses mud to build small pots in which to place its caterpillar prey.

Figure 42. A ground-nesting, hunting wasp of the genus *Ammophila* (Sphecidae).

Figure 43. A common paper wasp (*Polistes*, Vespidae), species of which are familiar to most home owners because they make the open-faced, paper nests frequently found hanging under house eaves. These wasps commonly prey on caterpillars but will kill other insects as well.

Figure 44. A female parasitoid wasp (possibly *Rubicundiella*, Ichneumonidae). These wasps lay their eggs in lepidopterous pupae (or prepupae).

Figure 45. Nesting blocks in a backyard habitat in Logan, Utah. The blocks have holes drilled in them, ranging from about 1/8 to 3/8 inch (3 to 9 mm) in diameter. The thickness of the block can vary, but 4 to 6 inches (10 to 15 cm) is about average. About 30 species of bees and wasps utilize these holes and cap them with resin, mud, or leaf material depending on the species. Photo by Frank Parker

Problematic Hymenoptera

Speaking of yellow jackets, hornets, and their ilk inevitably brings us to the subject of unwanted Hymenoptera, much as it pains me to use that phrase. Of all the tens of thousands of species of Hymenoptera, the social sorts are the most likely to cause any physical harm to the average gardener. Foremost among them are the stingers comprised of a very few wasps, the honey bee, and the red imported fire ant. The fewness of their species, however, is usually made up for in the potential vastness of their colony size and their relative quickness to anger. Of secondary concern are those few structural dis-engineers such as carpenter bees and carpenter ants. And finally come minimal annoyances such as ants in the pantry.

The potentially irritating and possibly dangerous Hymenoptera have been discussed earlier, so I won't repeat the problems they might cause. Instead we'll examine some ways to restrict or inhibit their occurrence. Again, I would like to emphasize that most of these creatures (even fire ants) offer potential benefits and their exclusion or elimination should only be undertaken when other alternatives are even worse.

In case of doubt as to how to handle an insect problem, it's always wise to consult a local extension agent, university brochure, or website for information about the insect in question. My own philosophy is to first do no damage, and so I generally consult a guide to natural insect control such as *Common-Sense Pest Control* (Olkowski et al. 1994), which provides many nonchemical options for problematic pests. In the end, however, I occasionally don't listen to my own advice and resort to chemical methods in an attempt to exert some control over a situation for which there is likely no control.

Singling out the most vicious offenders among the Hymenoptera first, I suppose that stinging (or predatory) wasps would win the contest. These all fall into the single group known as vespid wasps (Vespidae), which include yellow jackets, hornets, and paper wasps. Because colonies in colder regions are killed off in autumn, these wasps generally begin a new colony in the spring by a single female, namely the queen. Because of this, the gardener's best offense, if destruction is a must, is to simply eliminate the new queen, which is most easily done in the spring. In warmer regions, such as the southeastern United States, however, colonies may not die out in the winter and can become extremely large. As this generally happens in abandoned areas and not most gardens, the gardener need likely only be worried about new queens founding new colonies.

Of the three broad groups, yellow jackets, hornets, and paper wasps, the nest of the last is the most likely to be seen at such an early stage. Their open-faced, pendulous nests are frequently found under overhangs located around the home. Paper wasps are the least aggressive of all vespids, and if the single queen is eliminated, the problem comes to an end. Even if three of four wasps are at the nest they pose little problem. A wasp-killing insecticide works nearly instantly, or the nest could be encased in a bag (quickly of course) after dark and disposed of. During daylight hours, I have used a butterfly net to collect the few adults and then remove the nest.

Next come hornets, which usually build their nests so high up in trees that they seldom are seen, much less represent a threat. By the time hornet nests are seen they are generally so large that it is much too late for the ordinary gardener to do anything but accept them or call an exterminator. The ornery gardener might try chemical sprays, but it is still risky business to irritate these creatures. The introduced European hornet—a huge red, black, and yellow species—makes its nest in protected places such as attics and sheds. Much as with the paper wasps, if found early enough a small nest can be removed with some caution. I repeat: small nests. Large European hornet nests are again best left to the expert. An additional strategy is to screen off any entrances to open spaces in the house, garage, and sheds.

The in-ground nests of yellow jackets are likely to be rather large by the time they are discovered, by which time discretion is highly advised. These are, in my opinion, the most easily angered of the stinging wasps and the most irksome of the lot. Sprays designed specifically for the quick knock-down of wasps are available and work well for small colonies even in the daytime. Large colonies, however, are best approached at night and with caution. Huge colonies, which generally occur in warmer regions because of the absence of winter-kill, are totally outside the realm of amateurs.

Honey bees don't form a colony one bee at a time, as do the vespid wasps. The founding queen takes off in a swarm with thousands of followers and thus comes well equipped to occupy a new site when she finds one. Such swarms are attracted to hollow structures in which to establish a colony. Thus, hollow trees, poorly screened attics and sheds, and holes in an exterior wall through which they can gain access between wall studs will be welcomed by an invading bee swarm. Rarely, they simply construct an exposed nest in the crotch of a tree. The removal of honey bee colonies is not a job for the homeowner. With the invasion of the Africanized bee, this becomes doubly true as these are extremely aggressive and do not like moving objects near the entrance to their nest. If a honey bee swarm lands on or near your house or has already established a colony where it poses a threat, seek professional help at once. Don't think you are smarter than a honey bee. I'm sorry to say scarcely anyone is.

Ants can become problems in several different ways. In the home they may simply be a nuisance, whether feeding on our foodstuffs or feeding on the house (that is, carpenter ants). The main problem is that it is usually difficult to locate ants' nests, thus making it difficult to mechanically eliminate them. These sorts of ants can often be controlled with readily available baits, especially the kind that contains products that disrupt the physiological development of young (for example, juvenile hormone) in the colony. With these baits it is necessary to allow the ants to continue feeding long enough to take up the bait and deliver it back to their nest mates. I've had luck with these working on carpenter ants.

Garden ants of many different species are generally not much of a bother, but occasionally they build nests at the crowns of small plants, which are often killed in the process. More noxious in the extreme and serious in large parts of the southern United States are the red imported fire ant. I don't know of any way to mechanically prevent these problems, but several baits are recommended for ant control.

In the end, however, I suppose the easiest response to the question of how to eliminate these problematic Hymenoptera would be to do exactly the opposite of what we just explored in the first part of this chapter. You would eliminate all insect attractants of any kind. By providing no flowers or only those composed of highly double and pollenless forms, bees and many wasps would not be enticed to visit your garden. Clean up every twig, dead branch, and dead tree on site so there is no possibility of twig nesters gaining entrance. The draconian step of cementing over most of your soil would surely eliminate ground-nesting sites for solitary bees, predatory wasps, and most ants. And no water need be present as an attractant, either. In the end you could create a beautiful, if sterile, garden. No bugs, no birds, no nothing except the peace and quiet of a perfect paradise.

PART 2

THE LIVES OF HYMENOPTERA

HERE WE'LL EXAMINE the lives of Hymenoptera based on the five major biological groups, namely sawflies (including horntails and wood wasps), parasitic wasps (both true and stinging), predatory wasps, bees, and ants. It is somewhat tempting to simply list each family and present a paragraph about it, but many families have similar life histories and some families are rarely seen by entomologists, let alone gardeners. As a compromise, at the back of the book I present a table listing all the world's Hymenoptera families along with their larval feeding types and larval food sources. This is a bit academic, but it does provide a useful biological summary of each family without resorting to overkill. Also, the information only approximates reality. A parasitic family such as Ichneumonidae contains so many species that it is nearly impossible to characterize their biology beyond the fact that they are all presumably parasitoids of a broad range of other insects and spiders. The closely related family Braconidae is equally difficult to characterize because it has both a broad range of insect hosts and a few rarely encountered tropical species whose larvae feed in plant tissue. Even predatory spider wasps (Pompilidae), which all prey upon spiders, have species that act more like parasitoids than predators. The characterizations in the table of hymenopteran families, then, are generally useful for most species in the numerically larger families, and in some of the smaller families hold as nearly true as we can estimate at the present. Hymenoptera, however, always manage to lull us into a false sense of knowledge only to baffle us in the end.

In the chapters that follow, I summarize major aspects of each group, including factors of biological importance (with odd bits of interesting behavior), and how the gardener might interact with members of the group in his or her garden. In many cases, simply knowing a bit about the habits of a parasitic wasp, predatory wasp, or bee will aid in assuring the gardener that they are present even if they are never seen. For example, the mummified body of an aphid will confirm that aphid parasitoids are doing their job, or a scattering of sawdust near a broken plant stem may indicate the presence of a predatory wasp or bee in the process of constructing its nest. Biological processes swirl around us continually, and the more we know about the possibilities, the more likely we are to recognize and appreciate the causes.

5

The Garden's Cows: Sawflies

TECHNICALLY THIS GROUP is referred to as the Symphyta, or symphytans, but popularly it is given the umbrella term of *sawflies*. This terminology leads to some confusion because *sawfly* also refers to one
group within the Symphyta, namely the family Tenthredinidae. For
the purposes of our discussion we'll refer to this specific family as
common sawflies and retain sawflies as an all inclusive category. The
family Tenthredinidae is indeed common, as its members represent
two-thirds of all known members of the Symphyta. Contrary to their
popular name, sawflies are not flies (that is, Diptera). The name is
derived from their ovipositors, which are shaped like flattened saw
blades and are used to slice into plant tissue wherein they lay their
eggs. In this respect sawflies differ from other Hymenoptera in which
the ovipositor resembles a hypodermic needle.

It may come as a surprise to know that sawflies are credited with
being around since the birth of dinosaurs, that is, since the very early
Mesozoic (Triassic) period. In case you don't recall, this would mean
nearly 200 to 250 million years ago, give or take a few millennia. This
was the age of cycads, horsetails, mosses, ferns, conifers, and the first
inklings of flowering plants. As such, sawflies are considered the
most primitive Hymenoptera and are always placed first in any discussion of the order. Just as conveniently, the larvae of 99 percent of
all sawfly species are plant feeders and so serve as a good place to begin discussing garden-associated hymenopterans.

Numerically speaking sawflies are the smallest group of Hymenoptera, being represented in North America by 11 families and
about 1100 named species. Worldwide there are some 8200 species

and 2 additional families (ECatSym: Electronic World Catalog of Symphyta Online Version 2.0; see the list of useful websites), with the group being most numerous in arctic, subarctic, northern temperate, and humid subtropical and tropical regions (Smith 1993). Judging by the fact that only about one species in eight occurs in temperate climes, we temperate gardeners should consider ourselves lucky because the larval stages of these species are plant feeders (except the parasitic wood wasps, Orussidae). The majority of sawfly larvae are external plant feeders, so we are as likely, or more so, to encounter larvae than adults, unlike all the remaining hymenopterans to be discussed, in which the larval stage is hidden from view. In this respect sawflies are perhaps the least likely of all adult hymenopterans to be seen or recognized by the general public.

Adult sawflies feed principally on nectar, with the exception that most adults of one subfamily (Tenthredininae) of common sawflies are known to feed on small insects in order to obtain protein for egg development. For some reason, members of this subfamily are also considered exceptionally good pollinators. (Eventually I hope the reader will learn to accept the notion that when discussing the Hymenoptera there are always exceptions, and the Tenthredinidae, with its built-in diversity of world species, is no exception.) It might be possible to refer to members of the subfamily Tenthredininae as "predatory wasps," but as I stated early in this book, it is the larval stage that exemplifies the basic lifestyle of a hymenopteran, so we will stick with the notion that sawflies are plant feeders with the exception (as always) of the parasitic wood wasps.

Basically there are two types of larval plant feeders within the sawflies, the surface feeders that feed externally on a plant's leaves and the internal feeders that feed within a plant's tissues, including gall formers, leaf miners, fruit borers, and stem and trunk borers. The majority of sawfly larvae are external plant feeders, appearing much like those of butterflies and moths but having five to seven pairs of legs on the midsection of the body (that is, after the three pairs of thoracic legs), unlike caterpillars that have four or fewer pairs on the midsection. There are a few externally feeding sawfly larvae that appear slug-like (for example, roseslug, pearslug) and don't seem to fit this description, but they are in the minority. The internal feeders have at least three pairs of legs directly behind the head on the thorax but have mostly lost the remaining legs (or they are much reduced), so they do not resemble caterpillars. Whether external or internal feeders, sawfly larvae have minimally three pairs of legs and in this respect differ from the nondescript, legless larvae of virtually all the remaining Hymenoptera. Based on behavior, if the gardener finds a larva crawling about on a plant and it appears to be a typical "caterpillar," then it is almost assuredly either a butterfly/moth larva or that of a sawfly. The quantity of legs tells the difference.

In addition to the larval feeding habits of sawflies, adults also are unique in several respects centered around the way their body is put together. These aspects are technical and concern morphological features not found in the remaining groups of Hymenoptera. Adult sawflies have the abdomen broadly joined directly to the thorax, so there is no articulating (pivoting) ball and socket, unlike the remaining Hymenoptera. Sawflies look somewhat tubular or barrel-like, without any particular points of articulation except the head, wings, and legs. Another technical aspect of sawfly morphology is that each hind wing has three or more closed basal cells, but this is not a helpful distinguishing character when dealing with live or flying subjects.

The life cycle of most sawflies is fairly similar, with a few obligatory exceptions. Most adult sawflies are short-lived, appearing only long enough to mate, lay eggs, and die. Most often there is one generation per year, but some species may have from two to six. After feeding, most sawfly larvae fall to the ground and pupate in silken or cellophane-like cocoons in the soil or organic duff, and they overwinter (or hibernate) in the prepupal stage. Some species overwinter as eggs injected into plant tissue. A few species, such as the introduced pine sawfly (*Diprion similis*, Diprionidae), attach their cocoons to vegetation instead of pupating in the soil. Most of the stem borers pupate and overwinter within the stems they inhabit.

The easiest way to discuss sawflies would be based on the feeding habits of the larvae as external plant feeders, internal plant feeders, or rarely parasitoids. Unfortunately two of the families (Tenthredinidae and Xyelidae) are not exactly tidy when it comes to such things, so we'll discuss them first. This more neatly divides the remaining nine families into almost discrete biological categories, with a few minor feeding exceptions (of course).

Tenthredinidae (Common Sawflies) and Xyelidae (Xyelids)

Of the approximately 1100 total sawfly species known from North America, about 800 are members of the family Tenthredinidae. Elsewhere in the world there are nearly 5000 additional species. It likely goes without saying that such a large aggregation of species will not be entirely consistent in their feeding strategies, though we have established the fact that all are plant feeders as larvae. As common as they are, members of this family are not easy to characterize because they have no particular feature that lends itself to visual identification, which is technically based on antennal and wing vein characters. Adult tenthredinids are rarely more than ½ inch (15 mm) in length and come in various colors, most often black, with markings of red, yellow, green, blue, or white (Figure 46). Some are quite beautiful in life, but unfortunately they lose their coloration upon dying.

Although the vast majority of tenthredinid species are external foliage feeders as larvae, a few feed internally as gall formers (for example, on willow), a few are leaf rollers or leaf miners (such as the birch leafminer), and a few are borers in twigs (such as the roseborer) and fruits (cherry, apple, and pear fruit sawflies). Sawflies with slimy, slug-like larvae (such as the roseslug) are included in this family. Although tenthredinid larvae are relatively host specific, the total accounting of plant hosts for the family is both diverse and exhaustive, including just about any category imaginable from ferns, horsetails, rushes, sedges, grasses, shrubs, conifers, deciduous trees, and numerous herbaceous plants. As a result of host specificity, species are often named for the plants upon which they feed. For example, the bristly roseslug (*Cladius difformis*; Figure 47) feeds on rose leaves and may be characterized as a hairy slug. A hairless version, the pearslug (*Caliroa cerasi*), feeds principally on pear, cherry, and plum. Some of its cousins, the scarlet oak sawfly (*Caliroa quercuscoccineae*) and the European oakslug (*Caliroa annulipes*), skeletonize oak leaves. The raspberry sawfly (*Monophadnoides rubi*), feeds on raspberry, blackberry, and loganberry. A few other examples of host-named sawflies include: violet sawfly, birch leafminer, elm leafminer, larch sawfly, pear sawfly, gooseberry sawfly, willow leaf gall sawfly, and the

European currantworm. Among Hymenoptera, common sawflies are perhaps the most damaging to garden plants, though they are by no means commonly abundant.

The most likely garden variety common sawflies might well be the rose feeders, and the bristly roseslug is one of the most commonly encountered. It has many generations a year, thus being found throughout the growing season. The curled rose sawfly (*Allantus cinctus*) has two generations per year, one in spring and one in autumn, as does a European species with no common name (*Allantus viennensis*). The roseslug (*Endelomyia aethiops*) has but one generation per year. Another sawfly that primarily attacks wild rose and occasionally cultivated ones is the roseborer (*Ardis brunniventris*), the larvae of which bore and feed in shoots. Other garden variety sawflies include an occasional pest of blueberry that first feeds in buds with subsequent generations on leaves (*Neopareophora litura*); feeders on sensitive fern (*Hemitaxonus dubitatus*) and on bracken ferns (*Aneugmenus flavipes*); an introduced species with larvae that bore downward into stems of ferns (*Heptamelus ochroleucus*); and a feeder (*Phymatocera fumipennis*) on false Solomon's seal (*Smilacina*). Species of the related Eurasian family Blasticotomidae, with 13 species, are known to bore into the stems of fern.

With only a couple of dozen North American species and 150 worldwide, xyelid sawflies (Xyelidae; Figure 48) are a bit easier to comprehend than are the tenthredinids. Also they appear to be of little concern to the garden's well-being. This family is distinguished from other sawfly families by the third antennal segment, which is much longer than all the others combined. These are generally smallish wasps, about $1/8$ to $3/8$ inch (3 to 9 mm) in length, and often pale colored. Although the larvae of some species feed externally on the leaves of hickory and elm, most are internal feeders in the staminate flowers of pine or in developing fir shoots. These wasps are considered uncommon, but early one spring I once saw thousands of xyelids attracted to the male flowers of shrubby willows in full bloom. I haven't seen a xyelid since. Although insects often appear to be rare, we occasionally find examples of species rarity being more a factor of not being in the right place at the right time.

External Plant Feeders

The exclusively (or almost so) externally feeding sawflies are divided among the families Pergidae (pergids), Argidae (argids), Pamphiliidae (webspinning and leaf-rolling sawflies), Cimbicidae (cimbicids), and Diprionidae (conifer sawflies). In North America none of these families is particularly speciose, nor are they particularly destructive, with the exception of introduced conifer sawflies and the occasional outbreak of one or two garden irritants.

Pergid sawflies (Pergidae) are not well represented in North America, with species numbering only four. Larvae feed on oak and walnut. The family is identified in Canada and the United States largely by the number of antennal segments (six), but this is not likely to help even the most observant gardener. Not to worry, though, because not only are there few species, they seldom occur in abundance so they will rarely ever be seen. However, in other, largely tropical parts of the world, such as South America and Australia, there are more than 400 species (*Pergidae of the World*, see the list of useful websites). The life histories of most of these pergids are unknown, but existing records demonstrate that they are diverse feeders, attacking many plants including eucalyptus, ferns, potatoes, and even fungi. Some eat dead vegetation, a few are leaf miners or shoot borers, and in some species females exhibit maternal instincts, guarding their eggs or larvae (Schmidt and Smith 2008). Sometimes species in other countries potentially may be either harmful or even beneficial to our own gardens or environments if accidentally introduced. Such is the case with several pergids. A harmful pergid species, for example, might accidentally be introduced to California from Australia and have a field day consuming the vast quantities of eucalyptus found there. Or a South American species might be introduced that feeds on potato foliage. On the other hand, two pergid species are under consideration for introduction into Florida: *Lophyrotoma zonalis* (Figure 49) to control the noxious and invasive Australian paper bark tree (*Melaleuca quinquenervia*) and *Heteroperreyia hubrichi* (Figure 50) to control the Brazilian peppertree (*Schinus terebinthifolius*) from South America. The world being what it is, our best defense is to be familiar with the potentially invasive plants and insects that might reach our shores. Sometimes these discoveries are made in our own garden if we know about such things and are paying attention to the tiny creatures to be seen.

In North America there are about 70 species of argid sawflies (Argidae), with around 800 species worldwide. These wasps are distinct in having only three antennal segments, with the last (third) the longest in females and U-shaped in most males. Species are generally small, ranging from about 1/4 to 1/2 inch (6 to 12 mm), and are dark, often brightly bicolored red. North American argid species are leaf feeders, with wide-ranging hosts from deciduous trees (such as hazel, hawthorn, willow, birch, alder, oaks) to vines (including sweet potato and morning glory) and herbaceous perennials. One common species (*Atomacera decepta*; Figure 51) skeletonizes leaves of hibiscus, hollyhock, and other members of the Malvaceae. Argid larvae are gregarious (Figure 53) so that they may be found in large numbers, often causing damage before gardeners know what hit them (a good example of why the gardener should walk his or her garden on a regular basis). Another species (*Schizocerella pilicornis*; Figure 52) mines the leaves of portulaca, while a nearly identical species is an external feeder, though at one time both were considered to be a single variable species. Most wonderfully there is an eastern U.S. species (*Arge humeralis*) that can be counted as the gardener's best friend because its larvae consume the leaves of poison ivy.

The pamphiliids (Pamphiliidae), called webspinning sawflies (Figure 54), work a bit differently from other external feeders because they roll or tie leaves together, choosing to dine from within their protected hovel. Still, technically speaking they are dining externally relative to the leaves. These are mainly small, flattened wasps, ranging from 1/4 to 1/2 inch (6 to 12 mm), and are often black or dark with pale yellow or red markings. They have very thin, thread-like antennae, but don't really stand out from the crowd, so to speak, and are not terribly important as sawflies go. Pamphiliids number some 75 species in North America, with about 300 species throughout the rest of the world. Some species are gregarious, living in tent-like webs that might be confused with tent caterpillars. A few species, based on their names, might be considered pests if they ever became abundant, which they don't seem to do. These include the cherry webspinning sawfly (*Neurotoma fasciata*), plum webspinning sawfly

(*Neurotoma inconspicua*), peach sawfly (*Pamphilius persicum*), and blackberry sawfly (*Onycholyda luteicornis*; Figure 55). Most species of the family are conifer feeders or feed on leaves of deciduous trees deemed unimportant to human existence. A related family in Europe and Asia (Megalodontesidae), represented by about 40 species, consists of external feeders on many herbaceous plants species.

Among the most damaging of external plant feeders, the conifer sawflies (Diprionidae) would take first prize. They are considered enemies of the forest, and if you've ever had them in your garden you might know why. In my Maryland garden the redheaded pine sawfly (*Neodiprion lecontei*; Figure 56) commonly attacked several dwarf mugho pines. Their presence was initially noticed by the sudden disappearance of a foot's worth of terminal needles before the larvae were detected. The noted writer Cynthia Westcott, in *The Gardener's Bug Book* (1973), best described it when she wrote "The tiny young larvae . . . join together in gangs but they match the needles so well they are scarcely ever noticed until far too much damage has been done."

In North America there are nearly 50 species of conifer sawflies, with triple that number worldwide. Adult conifer sawflies are medium sized, about 1/2 inch (12 mm), and appear as pudgy, overweight little wasps. They differ from other sawflies in that the antennae are somewhat saw-shaped (females) or comb-shaped (males). Most, but not all, conifer sawflies begin life as gregarious feeders, with many tiny larvae feeding on needles of the terminal branches of their host. Eventually they disperse, but not before reducing the terminal stem to a naked stalk. This feeding can stunt growth or even kill trees, especially if they are young or not well established.

Conifer sawflies have proven most destructive in forest ecosystems, but Christmas tree farms, tree nurseries, and gardens also suffer outbreaks from time to time. As a result of potential and actual damage, some species have been the focus of extensive control programs. This is especially true of accidentally introduced species such as the European pine sawfly (*Neodiprion sertifer*), the European spruce sawfly (*Gilpinia hercyniae*), and the introduced pine sawfly (*Diprion similis*; Figure 57), all of which have proven worthy adversaries. The European pine sawfly was introduced into the United States in the 1920s. Its larvae prefer old foliage and may feed in clusters of up to 100 individuals, which migrate en masse as needles becomes depleted. This species differs from most sawflies in that larvae pupate in late summer, and adults emerge in autumn. Females then lay eggs, which overwinter in the current season's needles. The European spruce sawfly was also introduced to the United States in the 1920s but has a different life history. The larvae are solitary feeders on both old and new growth and overwinter in the prepupal stage in the ground. The pine sawfly was introduced to the United States from Europe in the 1910s. Its larvae are gregarious feeders on new or old foliage, becoming solitary as they mature, and overwintering takes place in the prepupal stage in the ground. If control is required in the garden, hand picking is recommended or a strong water jet can be used to dislodge the larvae.

North America has a considerable number of native conifer sawflies as well, but they are not considered as pesky as the introduced ones just mentioned. The diprionids have been given a surprisingly large number of common names relating to their primary hosts, or in some cases coloration. A few of the North American species include the arborvitae sawfly (*Monoctenus suffusus*), balsam fir sawfly (*Neodiprion abietis*), lodgepole sawfly (*N. burkei*), brownheaded jack pine sawfly (*N. dubiosus*), pinyon sawfly (*N. edulicolus*), blackheaded pine sawfly (*N. excitans*), redheaded pine sawfly (*N. lecontei*), slash pine sawfly (*N. merkeli*), white pine sawfly (*N. pinetum*), jack pine sawfly (*N. pratti*), loblolly pine sawfly (*N. taedae*), and hemlock sawfly (*N. tsugae*).

As undesirable as these conifer sawflies may appear to be, they do provide some benefits to the ecosystem. Birds and mice, for example, dine on overwintering pupae at a time when protein is scarce. Some gardeners may not see the benefits of mice, but they are considered one of the main biological agents controlling the European spruce sawfly and are equally likely to help control other conifer sawflies. Also mice form part of the greater food chain for other animals such as snakes and larger mammals. (For those who do not see the benefits of snakes, I apologize.) Conifer sawflies also serve as hosts for innumerable parasitic wasps and flies as well as fungi. An understanding of how biological systems work is often dismissed by humans who view the seemingly trivial components of the natural world as unimportant. Still, when dealing with the biological systems that sustain us all, it's probably best to step back and consider that not understanding may be preferable to suffering the consequences of misunderstanding.

Our final family of external plant feeders, the Cimbicidae (Figure 58), is certainly no match for diprionids when it comes to common names. What it lacks in recognition, however, it makes up for in size—especially the larvae. Larvae of all previous external feeders are dwarfed in size when compared to larval cimbicid sawflies, which can reach up to 2 inches (50 mm) in length. The adults are no slouches either, reaching up to $1^1/4$ inches (30 mm) in length, and in bulk and color they somewhat resemble a cross between a carpenter bee and a bumble bee. Unlike either bee or other sawflies, cimbicids have the tips of the antennae enlarged or club-shaped. There are only about a dozen species in North America, with nearly 200 found throughout the world. Although adult cimbicids are imposing in stature, they are nearly harmless unless handled. The tank-shaped body is muscular, and spines on the legs can impale careless fingers with an action that causes a fairly immediate release. The most common species of cimbicid to be found in North America is the elm sawfly (*Cimbex americana*), which feeds on elm and willow. Other cimbicid species feed on honeysuckle, snowberry, ash, maple, alder, poplar, willow, and birch. Adult cimbicids may girdle small twigs with their jaws, though this habit seems to be more accidental than necessary for their survival.

Cimbicids are rather uncommon as well as being large, so they make quite an impression on the psyche of a student of Hymenoptera. As a young entomologist I once chased a cimbicid through a dense undergrowth of willow along the overgrown banks of a creek. Heedlessly paying no attention to my own feet, but intense attention to the elusive, fast-flying cimbicid, I suddenly found myself laying face down in the bottom of a washout about 3 feet deep and 8 feet across. Fortunately it was simply a mud pit and not lined with rocks, so only my feelings were hurt. The saddest part was that I did not catch the cimbicid, nor did I seemingly learn any lessons about watching where my feet take me.

Internal Plant Feeders

The four families of internally feeding sawflies are Cephidae (stem borers), Xiphydriidae (wood wasps), Siricidae (horntails), and Anaxyelidae (cedar wood wasp). All these wasps may be referred to as *xylophagous*, a euphonious and seldom-used term meaning wood borers, though some actually bore in grass stems or bramble canes.

The stem sawflies (Cephidae) number about a dozen species in North America, with about 200 species worldwide. Boring in grass stems, bramble canes, or tree twigs, these are narrow-bodied, elongate wasps that range from ½ to ¾ inch (12 to 18 mm). They are generally black with yellow stripes or spots. The most widespread North American species (*Cephus cinctus*) attacks wheat stems and logically goes by the name of wheat stem sawfly. It attacks other grain species as well. Of more interest to gardeners would be the species that attack berries and roses. The raspberry horntail (*Hartigia cressonii*; Figure 59), for example, lays its eggs at the tips of raspberry, blackberry, loganberry, and even rose canes. The hatchling larva girdles the tip, causing it to wilt, then burrows downward into the stem's pithy core, sometimes reaching the roots. Larvae overwinter in the stems, then bore out as adults in the spring. The rose shoot sawfly (*Hartigia trimaculata*; Figure 60) has been reared from rose and blackberry, but likely attacks other hosts as well. The larva has the habit from time to time of girdling the stem above its feeding site. The dying stem then breaks off. The genus *Janus* has species with similar life histories but attacks shoots of willow (willow shoot sawfly, *J. abbreviatus*), oak (oak shoot sawfly, *J. quercusae*), and currants (currant stem girdler, *J. integer*; Figure 61). Generally, when a plant is not suffering from water stress and individual growing tips start to appear wilted, it is a sure sign that some sort of boring insect is at work.

The wood wasps (Xiphydriidae) number 10 species in North America, with 138 worldwide. Wood wasps range from about ⅜ to ¾ inch (9 to 18 mm) in length and appear like small versions of the horntails. Larvae of wood wasps bore into the dead and decaying limbs and branches of deciduous trees, including maple, hickory, birch, and linden. This is unusual in that most wood-boring sawflies prefer living tissue. In at least some species it has been discovered that the larvae are feeding either upon symbiotic fungi that the adult female injects as hyphal fragments at the time of egg deposition or upon the wood softened by the fungal growth. When female larvae (but not males) pupate and become adults, they store fungal hyphal fragments from their burrow in pockets associated with the ovipositor so that they can take an incipient fungi starter with them to pass on to their offspring.

The horntails (Siricidae) are undoubtedly the largest of the sawflies, some reaching up to 2 inches (50 mm) in length. They are among the largest of all Hymenoptera as well. Typical coloration is black with yellow bands on the abdomen and some red on the thorax. Some species are entirely metallic dark blue. Females of all species have a horny plate projecting from the tip of the abdomen over a portion of the ovipositor, and males have a slight angular terminal projection. As formidable as these wasps appear, they cannot sting or even poke. The larvae also have a terminal spine. There are about 30 species in North America and about 125 worldwide. As with the wood wasps, females of all but one genus carry symbiotic, wood-rotting fungal hyphal segments that they inject along with their eggs into dead and dying coniferous or deciduous trees. These wasps normally attack trees under stress or that are dying from other causes, and horntails are frequently found after forest fires, where they will attack both vertical and felled trees. The larvae tunnel either in sapwood or heartwood, in which they are aided by the fungus, which softens up the wood and makes it possible for the larvae to feed, though it's not quite certain whether they feed on the wood, the fungus, or both. Pupation takes place just under the bark. Horntails attack a large number of tree species, but depending upon the subfamily they attack either deciduous trees or conifers such as cypress, redwood, pine, Douglas-fir, larch, hemlock, and spruce. Although horntails will not lay eggs in finished wood (that is, lumber), their larval cycle is long enough (1 to 3+ years) that mature adults are known to emerge from finished lumber that has not been kiln dried.

The pigeon tremex (*Tremex columba*; Figure 62) is a North American horntail that oviposits in several species of hardwoods, including silver maple, ash, cottonwood, beech, oak, apple, pear, sycamore, hackberry, and elm. Its preference is for maple and beech, but sometimes wood is wood and preferences can be overcome. This species injects its eggs up to ³/₄ inch (18 mm) into solid wood. With the aid of the wood-softening fungus, larvae of the pigeon tremex can tunnel anywhere from 6 inches (15 cm) to nearly 9.5 feet (3 m). A similar appearing species, doubly named the yellow-horned horntail (*Urocerus gigas flavicornis*), also occurs in North America, but the horn at the tip of its abdomen is much longer than that of the pigeon tremex.

Finally, we reach the last of the internally feeding sawflies. The family Anaxyelidae is known from but a single living species, *Syntexis libocedri* (Figure 63), the cedar wood wasp. This so-called living fossil is the remaining member of a family represented by 34 fossil species reported from China and Russia. The placement of this group has been a puzzle, and over time they have been placed in either the family of stem sawflies or of wood wasps, but now they come to reside in their own family. *Syntexis libocedri* attacks recently burned incense cedar, red cedar, and juniper in the western United States and Canada. Females may even lay their eggs into smoldering or recently charred stems, and the larvae tunnel into them. Adults are black (not from being scorched, by the way) and about ¹/₄ to ³/₄ inch (6 to 18 mm) in length. If cedar wood wasps were a serious threat to our gardens, the obvious means of control would be to prevent your trees from catching fire.

Parasitic Wood Wasps (the Exception)

This category concerns only one family of sawflies, the Orussidae, or parasitic wood wasps, of which fewer than 10 species are known in North America and fewer than 100 in the world. North American species are small, about ¹/₄ to ¹/₂ inch (6 to 12 mm) in length, and it is unlikely that a gardener will ever see one of these wasps, so they are mentioned simply for the sake of completeness. These are enigmatic wasps full of bizarre morphological adaptations, including an ovipositor twice the wasp's length entirely concealed within the body like a twisted lasso. Before being laid, the eggs are said to be longer than the body itself but are folded up internally as is the ovipositor. There are many fossil examples of these wasps dating back nearly 100 million years, and the group is considered to be intermediate (a "missing link") between the sawfly group (Symphyta) and the remainder of the Hymenoptera treated in the following chapters. Orussids are found inside plant stems, but they don't feed on plant tissue. Instead they seem to attack the larvae of wood-boring plant feeders including beetle larvae and probably larvae of their own distant relatives, the horntails. I have never encountered orussids in the field, but I've read that they can be common within restricted areas. For this, I will have to take the author's words. And now you know as much about these rare and mysterious wasps as I do.

6

The Garden's Police:
True Parasitoids, Stinging
Parasitoids, and a Few Surprises

AMONG THOSE WHO STUDY Hymenoptera, the term *parasite* is not routinely used for wasps that indirectly attack other insects in a sneaky and perverse fashion. The preferred term is *parasitoid(s)*, a group in which I have specialized my entire adult life. Entomologists rarely, however, call them parasitoidic wasps, preferring instead to stick with parasitic wasps because it's easier to say—which makes little etymological sense, I agree.

As defined strictly on technical morphological grounds, parasitoids of the suborder Parasitica (from the Latin *parasitus*, a sponger or parasite) are an assemblage of some 47 worldwide families defined by an ovipositor constructed for laying eggs. Unfortunately for tidy people, there are another 15 families that have parasitic habits but are classified in the suborder Aculeata (from the Latin *aculeus*, to sting) because their ovipositor is constructed for stinging not egg laying. Parasitoids of this latter group may even move their host, once stung, to a protected place, quite unlike a typical parasitoid, and in this respect approach the nesting behavior of hunting wasps, which are related members of their suborder. Because I chose to discuss groups by biology rather than morphology, the problem arises as to what to call the different subsets of parasitoids to better discuss their behavior. In this chapter I am going to call them *true parasitoids* and *stinging parasitoids*, with a collective term of *parasitoid(s)* for both.

Parasitoids are among the least well known insects never seen or recognized by the public. Numbering nearly 89,000 described species, these wasps may be found throughout the hospitable parts of the environment including fresh water and even floating through the air. Because most are relatively small, difficult to identify, and innocuous in the apparent (but not real) scheme of things, it is likely there may be 5 to 10 times as many of these creatures awaiting discovery.

True Parasitoids

The vast majority of true parasitoids attack other insects, but a relatively few attack plants. Forty-seven families and 76,000 species of Hymenoptera are true parasitoids (see Parasitica, table of hymenopteran families). Of these only about 2900 are plant feeders, mostly in the gall wasp family Cynipidae. A quick glance at the hosts listed in the table of hymenopteran families might give you some idea of the tremendous range of insect (and plant) hosts that fall victim to these wasps. Numerically the two largest superfamilies of parasitoids are the Ichneumonoidea (families Ichneumonidae, Braconidae) and the Chalcidoidea (many families). Between these two, few insects remain that are not confronted by an imminent and gruesome threat of death. The superfamily Chalcidoidea alone is known to attack thousands of host species in 14 insect orders including Blattaria (cockroach eggs), Coleoptera (beetles), Diptera (flies), Hemiptera (true bugs, scales, mealybugs), other Hymenoptera, Lepidoptera (moths, butterflies), Mantodea (mantid eggs), Neuroptera (lacewings), Odonata (dragonfly eggs), Orthoptera (grasshopper, cricket eggs), Psocoptera (barklice eggs), Siphonaptera (flea larvae), Strepsiptera (twisted-wings), and Thysanoptera (thrips), as well as members of the class Arachnida: spider eggs (Araneae), ticks, gall-forming mites (Acari), and cocoons of pseudoscorpions (Pseudoscorpionida). They even jump phyla into the Nematoda, attacking gall-forming nematodes. In addition, members of this superfamily (and rarely Ichneumonoidea) have strayed off the evolutionary pathway, so to speak, becoming specialized seed feeders, gall formers, and obligate fig pollinators in the process—thus, the "surprises."

It may seem incongruous to include plant feeders among the parasitoids. After all, sawflies are the acknowledged plant-feeding champions, so why not put parasitoids with that group? That might be a biological grouping of convenience, but then we would have to add all the bees (pollen feeders), some ants (fungus and seed feeders), and some predatory wasps (the pollen wasps, subfamily Masarinae). So plant feeding is not entirely a good grouping character. The problem would become even more complex if we tried to group all Hymenoptera by feeding behavior alone because there are some sawfly parasitoids, some predatory wasps that act like parasitoids, some parasitoids that act as predators, and even some bees and ants that are social parasites. And finally, when considering parasitoids as an assemblage, relatively few groups have gone over to plants and these are always found within larger, truly parasitic groups to which they obviously belong and from which they have obviously simply traded insect hosts for plant hosts. For instance, the genus *Megastigmus* (Torymidae) is a group of about 150 species in a family of more than 900 species of typical parasitoids, but members of this one genus may be either parasitoids of gall formers, some may be gall formers (though this is not yet known for certain), and many are definitely internal seed feeders.

Parasitoids represent a numerically huge and relatively difficult area of Hymenoptera to cover with any degree of completeness. As the large number of species and families suggests, there is a complex degree of morphological as well as quantitative diversity to be found even within the same family. The smallest species of all known insects is a parasitic wasp of the delicately named fairyfly family (Mymaridae) having the hopelessly unpronounceable name of *Dicopomorpha echmepterygis*. Blind and wingless males of this species measure 0.005 inch (0.139 mm) in length, which is slightly longer than the width of a human hair (0.004 inch, 0.1 mm; Gahlhoff 1998). Females of this fairyfly lay two to four eggs inside the egg of its host, a barklouse (Psocoptera), the adults of which are less than 3/16 inch (5 mm) in length. Small as this wasp might be, "males smaller than those of *D. echmepterygis* may exist among parasitic wasps, especially those that parasitize eggs of other insects" (Gahlhoff 1998). At the opposite extreme among parasitoids is an ichneumon wasp (*Megarhyssa atrata*, Ichneumonidae) that reaches up to 7 inches (17.5 cm) in length mostly due to the greatly elongated ovipositor. These wasps attack horntail larvae (that is, sawflies) living in tree trunks and so must be able to reach the host. In general, there is a fairly direct correlation between the size of a wasp relative to that of its host, but exceptions exist.

Because parasitoids represent a numerically huge group of species, I have not approached them as was done for the sawflies. Instead of discussing each family as a unit, we'll concentrate on the way they work and their types of biology as a way of better understanding what these wasps are all about. Parasitoids attack other living things, but different species do so within a given set of parameters. The remarkable aspect of parasitoid behavior is that most species—though not all—are fairly specific in the host species, host stage, and strategy they use. A species that attacks a moth larva, for example, will generally not attack a beetle larva; a parasite of eggs will not attack larvae. As a result, the range of hosts attacked by a single family of parasitoids may be huge, but species, groups of species, and even genera within that family may specialize upon just a few types of host. A few families have an extremely limited host range, but they are not common. Eucharitidae, for example, attack ant pupae and only ant pupae, whereas Evaniidae attack only cockroach eggs.

In the following two sections, I treat those parasitoids that attack other insects and spiders and then those that are plant-feeding species. Rather than giving a horrifically detailed account of the highly specialized interactions between parasitoid and host, I simply give examples of behavior based on the stages of host attacked as illustrations of what is happening somewhere in your garden, even if you don't believe it.

Parasitoids as Carnivores

Here I describe the basic biological modes used by parasitoids that attack other insects and spiders, as well as a few unusual hosts. Simplistically they may be thought of as wasps that attack either the egg, larval, pupal, or adult stage of their host, though this is way too simple, as we shall see when we examine each stage. These basic modes of attack may be accompanied and complicated by several basic feeding strategies, such as primary or secondary (hyperparasitoids) feeders, internal (endoparasitoids) or external (ectoparasitoids) feeders, permanent or temporary paralyzers, and solitary or gregarious feeders. Thus, there are many combinations and permutations of feeding modes, but we will stick with the basics and mention only a few parasitoids that bend the rules just a bit.

Egg parasitoids

An egg, sitting still as it does without the ability to move or fight back, makes an easy target for a wasp intent on finding a host for its own offspring. The egg is a protected bag of nutrition just waiting to be exploited, but it requires a high degree of specialization for any parasitoid to make use of. Insect eggs are small, and any parasitoid that develops inside an egg would need be even tinier. Imagine living in the egg of a thrips (Thysanoptera), itself barely visible to the naked eye. Some insect eggs can support two, three, or even more parasitoids. as we saw above in the example of the fairyfly in the barklouse egg. Before a parasitoid is able to complete development inside the confines of an egg—a mean feat by any standard—its mother must first find the correct host species of egg in which to oviposit. Obviously these egg parasitoids have achieved this goal or they would not exist, and their numbers are legion, with more than 5000 described worldwide.

Two families and one subfamily of parasitoids exclusively complete development entirely within an egg. Trichogrammatidae primarily attack eggs of true bugs and leafhoppers (Hemiptera) and moths and butterflies (Lepidoptera) and less commonly beetles (Coleoptera), thrips (Thysanoptera), flies (Diptera), and dobsonflies (Neuroptera). The eggs of Trichogrammatidae, sometimes called trichos, are offered for sale to control moth and butterfly larvae, via the egg stage. At best they are a cursory form of birth control and do not have a visible or immediate impact upon either larval or adult populations. Species of the family Mymaridae are primarily parasitoids of leafhoppers and their relatives, but they also are known to attack barklice (Psocoptera), beetles, flies, and dragonflies and damselflies (Odonata). Species of the subfamily Scelioninae (Platygastridae, Figures 64, 65) attack eggs of grasshoppers and crickets (Orthoptera), mantids (Mantodea), webspinners (Embiidina), true bugs, lacewings (Neuroptera), beetles, flies, butterflies and moths, and spiders (Araneae).

Several species in each of these families are capable of reaching eggs underwater either by walking or swimming to them. One species of mymarid swims using its wings to reach the eggs of a predaceous diving beetle (Dytiscidae), even mating inside the egg (Bennett 2008). As an adult this mymarid can survive several days underwater.

A species of trichogrammatid swims using its legs and parasitizes diving beetle eggs and those of dragonflies and damselflies (Odonata). Up to 70 adult wasps have been reported emerging from a single dytiscid egg. A species of scelionid wasps lays its eggs in those of water striders (Gerridae, Hemiptera; Askew 1971).

Another mode of attack is used by egg parasitoids, but it differs substantially from the internal approach. In this case the parasitoid lays its egg(s) amid host eggs in a cluster. These clustered eggs may be in a mantid or cockroach egg case, in a spider egg sac, or simply exposed, as in a cluster of moth eggs on a leaf. In this case the parasitoid larva actually acts as a predator because it actively dines progressively from one egg to another. There is but a single family of wasp whose larvae feed exclusively in this manner, namely the ensign wasps (Evaniidae; Figure 66), so called because their abdomen is flag-shaped. Members of this family feed upon eggs inside cockroach egg cases. Unlike the internal egg feeders, ensign wasps are not constrained by the diameter of a single egg and may attain a size large enough to be seen, 1/4 to 3/8 inch (6 to 9 mm) in length. On occasion, wasps can signal events unsuspected by the ordinary viewer, so if you see these easily identified wasps flying about in public buildings, as I have, you know that cockroaches are not far away.

Because the predatory form of egg development is much easier than squeezing into an egg itself, this form of predatory parasitism has developed sporadically within several families. It is the most common form among the relatively few parasitoids that attack spider egg sacs—including those of the black widow—as well as mantid egg cases. Such families include Ichneumonidae, Encyrtidae, Eurytomidae, Eupelmidae, Eulophidae, Pteromalidae, and Torymidae.

Egg-larval/egg-pupal parasitoids

In some instances a parasitoid lays its egg within the egg of a host, but it remains dormant and does not hatch until the host has developed into a larva or pupa. At some point the parasitoid egg hatches and consumes the larva or pupa from the inside. The egg-larval strategy is relatively uncommon among Hymenoptera in general, but one large subfamily of Braconidae (Cheloninae) has adapted the lifestyle with zeal. This subfamily primarily parasitizes caterpillars (Lepidoptera) concealed in plant stems, buds, fruit, and leaf rolls. The egg-larval habit is also found scattered sparingly in a few species of Ichneumonidae, Encyrtidae, Torymidae, and Eulophidae. The most extreme case is shown by the encyrtid species *Copidosoma floridanum*, in which a female lays one or more eggs into the egg of a host moth. After the caterpillar develops, thousands of wasp embryos emerge from each single tiny egg. In the end, the host larva is a mere sausage casing containing thousands of wasps that explode from its lifeless carcass. At a record of more than 3000 parasitoids in one host, this wasp is cited in the *University of Florida Book of Insect Records* as having the largest parasitoid brood of any insect (Alvarez 1997).

In some cases the parasitoid larva waits for the host larva to enter its prepupal or pupal stage before emerging, in which case it is an egg-pupal parasitoid. The difference between egg-larval and egg-pupal is not always entirely clear, but it seems that the egg-pupal strategy is less common than the egg-larval approach. The subfamily Platygastrinae (Platygastridae, Figure 67) is credited largely as an egg-pupal parasitoid, but apparently some have also been reared as egg-larval species. In either case, they are all specialist parasitoids of gall midges (Cecidomyiidae, Diptera). The egg-pupal habit pops up sporadically within other families, notably ones that also show egg-larval parasitism such as Braconidae, Ichneumonidae, and Encyrtidae.

Larval parasitoids

Although internal egg parasitism has been adopted by a few families of wasps, it does not seem to have been as favorable an option as feeding upon a nice, fat, juicy insect larva. If one had to pick a single host category from among all available choices, parasitoids that attack the larval stage of its host would win ovipositor down. It is the most common characteristic among nearly all parasitoid families except those exclusively devoted to internal egg parasitism and the few plant-feeding families. Of all the parasitic wasp types known to gardeners, larval parasitoids are likely the ones that first come to mind, including particularly the white cocoons that show up on the back of tomato hornworms.

Larvae are the second immature stage of insects that have a complete metamorphosis, that is, with four different life stages: egg, larva, pupa, and adult. The insects included in this group are flies (Diptera), beetles (Coleoptera), butterflies and moths (Lepidoptera), lacewings and antlions (Neuroptera), twisted-winged parasites (Strepsiptera), scorpionflies (Mecoptera), caddisflies (Trichoptera), fleas (Siphonaptera), and Hymenoptera. Larvae of all these orders, even those of caddisflies and moths that live underwater, are attacked by hymenopterous parasitoids.

Unlike the stationary bag of nutrition found in an egg, a larva poses several obstacles to a parasitoid. At one extreme are the exposed, easily discovered, plant-feeding larvae that are able to resist being parasitized by running, falling, head butting, back arching, vomiting, twitching, or the like. At the other extreme are the many larvae engulfed in plant tissue such as tree trunks, galls, stems, mines, and leaf rolls. Some, such as beetle grubs, are buried in the soil. These latter hosts are relatively immobile. As can be seen from the table of hymenopteran families, many parasitoids are associated with larvae of all kinds, so we'll limit this discussion to a few different sorts of hosts to demonstrate the diversity of how larval parasitoids get the job done.

Among parasitoids, quite a number of uncommon families are associated exclusively either with wood-boring beetle larvae (such as Vanhorniidae), wood-boring horntail larvae (such as Ibaliidae), or wood-nesting wasp or bee larvae (such as Gasteruptiidae; Figure 68). A few families attack both beetle and horntail larvae (such as Aulacidae, Stephanidae), and then there are large families such as Ichneumonidae and Braconidae, each of which contains many species that attack all sorts of wood-boring insects. Because wood-boring larvae are cryptically located under bark or inside woody twigs and stems, a female wasp must first locate the hidden target. She does this by using her antenna to sense movement within the substrate—much as a doctor uses a stethoscope to hear a heart. Once located, the female wasp must then decide where to insert her oviposit to make a direct hit on the precise target area in which to lay her egg. Because beetle or horntail larvae are located in tree trunks, a wasp must have a long ovipositor to reach them, but amazingly the ovipositor does not have to be strong as a nail. It is relatively flexible and, given time, it will saw its way deep into plant tissue by use of tiny serrations at its tip.

The larvae of moths and butterflies are more familiar to gardeners than are the wood-boring sorts, because they are external feeders and often cause easily visible damage to plants. Common examples are the tomato and/or tobacco hornworms (*Manduca*), both of which feed on tomato and are of huge size and vast appetite. We occasionally see such a larva with white, cottony capsules on its back. Some mistake these for eggs, but they are the cocoons of a braconid wasp, usually *Cotesia congregata* (Braconidae; Figure 69, right), that has previously parasitized the host larva. These wasps lay a batch of eggs within the body of an early-stage caterpillar. Along with the eggs a virus is inserted that eventually arrests development so that the host does not pupate when it reaches its last stage. At this point the wasp larvae that have been consuming it alive from the inside burrow through its skin and pupate on the body of the host (Figure 69, left).

There are endless varieties of insect larval stages as well as parasitoids that attack them, and I've only mentioned two common sorts. I'll mention a few less common parasitoids and their hosts simply to point out what takes place in the world we seldom think about. In Europe, there is a wasp (*Bairamlia*, Pteromalidae) that parasitizes the larvae of fleas (Siphonaptera), likely a surprise to most readers simply because few know that fleas have a larval stage. Greatly elongate pelecinid wasps (Pelecinidae), up to $2^3/4$ inches (70 mm) in length, have substituted a long, flexible abdomen for an ovipositor, sticking the entire abdomen into the soil to reach underground scarab beetle grubs (Scarabaeidae). Ibaliid wasps (Ibaliidae) search for the ovipositor hole made by a horntail wasp (Siricidae) in tree trunks, then lays its own egg down the same hole. The ibaliid searches for holes based on fungus that is injected during oviposition by the horntail.

Things can become a bit more complicated and dangerous when considering the various levels of parasitism. For example charipid wasps (Charipinae) are specialist hyperparasitoids, some species attacking braconid wasp larvae (Braconidae) inside aphids (Aphididae) and others attacking encyrtid (Encyrtidae) larvae inside jumping plant lice (Psyllidae). Trigonalid wasps (Trigonalidae), lay large numbers of eggs on plant foliage that must be eaten by caterpillars or sawfly larvae as they feed. Once hatched inside the host, the trigonalid larva will only feed on the larva of a parasite already inside the caterpillar. Other parasitoids are more direct and fearless. Adults of the heavily armored chalcidid wasp (*Lasiochalcidia*, Chalcididae) walk right into the ferocious jaws of an antlion larva, which clamp around the wasp without harming it. Meanwhile, as the antlion holds the wasp in what would appear to be a deadly headlock, the wasp is injecting an egg into the antlion's throat.

There are many such remarkable and unlikely instances of parasitoid behavior, including species that attack their prey underwater. In North America, some species of the large family Ichneumonidae are aquatic, parasitizing either moth larvae or caddisfly larvae living on or under rocks in moving streams. In the case of caddisflies, the wasp lays its eggs inside the larval case. The wasp larva consumes the host, pupates, then overwinters in the case as an adult. It emerges the following summer. It likely comes as more of a surprise that some moth larvae live underwater than it does that some parasitoid has managed to find it. In Australia, there are quite normal female ichneumonids that parasitize moth larvae, while their male counterparts are off fornicating with orchids. The orchid, it seems, produces the same mating chemical (pheromone) as the female wasp, thus luring unsuspecting males to pollinate it. Thus, once the male has helped its own kind reproduce, it also helps out orchidkind. As with many animal species, when blinded by lust males are often not quite as bright as they might be.

Larval-pupal parasitoids
As with parasitoids that attack eggs but develop within the larvae, some parasitoids attack larvae but emerge from the pupae. For some reason flies (Diptera) seem to be favored hosts among these wasps. The family Figitidae (Figure 70) has several specialist subfamilies (Figitinae, Eucoilinae) with large numbers of species that attack fly larvae, especially those associated with dung. Because the wasps emerge from the pupal stage, their true technique of parasitism is easily misinterpreted. Braconidae is another family that contains larval-pupal parasitoids (Opiinae, Alysiinae)—flies again, but primarily stem- and leaf-mining flies and fruit-infesting flies. Other families included in this biological category include Ichneumonidae, Chalcididae, Eulophidae, Pteromalidae, Encyrtidae, and Eurytomidae.

Pupal parasitoid
The pupal stage of insects is another inactive life stage that we might expect to be heavily attacked by parasitoids. Parasitizing the eggs or

larval stage of a host and emerging from the pupae appears to be more common than directly attacking pupae, perhaps because they are a bit thicker skinned than larvae. Still, a pupa is a large target just waiting to be used, and so some wasps chose to do so.

We saw above that many species of the family Figitidae are larval-pupal parasitoids of flies. The subfamily Aspiceratinae (Figitidae) dispenses with the larval stage and goes directly to the pupa. Unlike its cousins that attack dung-inhabiting flies, the aspiceratines unfortunately specialize in flower fly pupae (Syrphidae). I say "unfortunately" because flower fly larvae are great aphid predators in our gardens. Strict pupal parasitoids are found in the families Braconidae, Ichneumonidae, Pteromalidae, Chalcididae, Torymidae, Eulophidae, and probably others, though it is difficult to know if the pupal stage was parasitized or the larval, which likely leads to many false records in the literature (Figure 71).

Nymphal parasitoids

Nymphs are the immature stages of insects that undergo simple metamorphosis, that is, insects that only have three life stages: egg, nymph, and adult. Common insects included in this group are grasshoppers and crickets (Orthoptera); mantids (Mantodea); dragonflies (Odonata); thrips (Thysanoptera); and true bugs, scales, mealybugs, and psyllids (Hemiptera). Nymphs are essentially small, wingless replicas of the adult to come and are capable of rapid movement including running, jumping, and swimming (in the case of dragonflies). Most nymphs are not attacked by parasitoids, but for some reason those of the order Hemiptera seem to be highly favored, and certain groups of parasitoids take advantage of these immatures with great delight. The most common and important hosts include several families well known to the gardener, such as aphids (Aphididae), plant bugs (Miridae), lace bugs (Tingidae), stink bugs (Pentatomidae), whiteflies (Aleyrodidae), armored scales (Diaspididae), soft scales (Coccidae), and mealybugs (Pseudococcidae).

Aphids, plant bugs, lace bugs, and stink bugs are a bit different from the others in that the nymphal stages are free moving. Few of us have ever seen an aphid that could outrun a parasitoid, but the others are quite capable of quick movement. All four groups have succumbed as hosts to members of the wasp family Braconidae. Aphids are the favorite game of one group of braconid wasps called aphidiids (placed in their own subfamily Aphidiinae; Figure 72, left). Both nymphs and adults are attacked by aphidiids, and the end product is a motionless, mummified aphid appearing much like a small brown paper aphid replica (Figure 72, right). If aphids in the garden are left unsprayed long enough, it is almost certain these wasps will find them. I have been surprised to find hundreds of such wasps in my greenhouse in late winter. I have no idea where they came from. There are somewhere around 100 described species of aphidiids in the Western Hemisphere. Many other bug nymphs fall victim to parasitoids of the braconid subfamily Euphorinae, which we will reencounter when we reach the adult stages attacked by parasitoids.

Whiteflies have a quiescent preadult stage (somewhat resembling a pupa), whereas scales and mealybugs have a barely mobile crawler stage that soon becomes immobile when it settles down to feed. Both groups are sitting ducks, so to speak, for any parasitoid that comes along. And there are plenty of wasp species that do so, all in the chalcidoid families Aphelinidae, Encyrtidae, and Signiphoridae. Many of these wasps are extremely tiny, from $1/16$ to $1/8$ inch (1.5 to 3 mm), so they are rarely seen by the gardener.

Adult parasitoids

Adult insects, as you might imagine, are generally much more mobile than larvae, are more heavily armored, and can be considerably more belligerent, a fact reflected in their rare use as hosts by parasitoids. It is here that the ferocious predatory wasps take command of the playing field, but we'll examine that aspect in the next chapter. Among the relatively uncommon cases of adults being used as parasitoid hosts, the prize for bravery goes to a group of ichneumonid wasps (Polysphinctini, Ichneumonidae) that have the brazen ability to attack adult orb weaving spiders in their webs. The spider is stung and temporarily paralyzed by the wasp, which then lays an egg externally on its abdomen. The spider eventually awakens and continues life unaware that a larval parasitoid is draining away its body fluids until it dies. Then the parasitoid has the audacity to pupate and spin a cocoon attached directly to the spider's web.

The prize for diversity of adult insects attacked by parasitoids goes to the related family Braconidae, which has a couple of subfamilies that display this unique habit. The subfamily Euphorinae most commonly seeks out adult beetle hosts, including leaf beetles (Chrysomelidae), weevils (Curculionidae), bark beetles (Scolytinae), and lady beetles (Coccinellidae). The next most common group of hosts is true bugs (Hemiptera), including mirid bugs (Miridae), lace bugs (Tingidae), and stink bugs (Pentatomidae). Because the adults and nymphs of these bugs are not terribly different, the nymphs are also attacked. Less commonly used as hosts are adult lacewings (Chrysopidae), adult and nymphal barklice (Psocoptera), adult ichneumonids (Ichneumonidae), and adult bees (of several families). Another subfamily of braconids (Neoneurinae) has been reared from adult ants (Formicidae). Rarely a species or two of some other families attack adult insects, but they are of little consequence compared to these braconid subfamilies, which specialize in taking on the grownups.

Noninsect parasitoids

Hymenopterous parasitoids have an overwhelming involvement with nearly all orders of insects (class Insecta). Careful readers also will have noticed that spiders (class Arachnida, order Araneae), which are not insects, are on the receiving end of some of these parasitoid attacks as well. Once beyond these two groups of hosts, we run into a few more surprising noninsect odds and ends that are attacked by parasitoids. Ticks (Arachnida, Acari), for example, are attacked by a few species of encyrtid wasps (Encyrtidae). The wasp lays its egg inside the abdomen of a tick, but development of its larva is delayed until the tick becomes engorged with blood from feeding on its host. Another odd host is mites (Arachnida, Acari), which are so tiny it seems nothing would bother them. Gall-forming mites, however, create a mite-filled ball of nutrients that is attacked by at least some species in the family Eulophidae. Eggs of these eulophid wasps are laid in the gall, and the larvae act as predators grazing among the defenseless mites. In Australia a species of eupelmid wasp (Eupelmidae) attacks pseudoscorpions (Arachnida, order Pseudoscorpionida), which are barely more than 1/4 inch (6 mm) in length. In Africa a species of eulophid parasitoid is known to parasitize a host completely outside the phylum of insects and spiders (Arthropoda). It attacks a gall-forming nematode (phylum Nematoda, order Monhysterida).

Parasitoids as Plant Feeders: Surprises

Given the extreme range and diversity of animal hosts, as well as the intricacies of life stages attacked, it should come as little surprise that parasitoids have also conquered the realm of plants. Included in this section are comments about relatively common, but not necessarily commonly seen, parasitic wasps that feed on plant tissue. The easiest to define are those wasps that feed internally in seeds and those that feed internally in plant tissue, causing it to swell up and produce galls. The not so easily defined are the fig wasps about which misinformation abounds. They have been given a category unto themselves. Then, too, are the various combinations of insect-eating and plant-eating interactions about which we are just barely aware.

Some confusion surrounds plant-feeding parasitoids, stemming mostly from our profoundly poor knowledge of what actually goes on within the life cycle of these minute wasps. For example, when a wasp is reared from a seed, it is entirely possible that it was parasitizing some other creature that is actually feeding on the seed. Galls are an even greater resource for confusion as there are sometimes other wasp guests (inquilines) living in the gall tissue along with the gall former. When a wasp is reared from a gall it could be the gall former, it could be a guest in the gall tissue, or it could be a parasitoid of either. Even more complicating is that some parasitoid larvae are known to attack and eat the gall-forming larva, then finish development by feeding on the gall itself. Others are known to eat the gall tissue until the gall former grows up, then it becomes dinner.

Plant-feeding parasitoids are found in relatively few families compared to the hordes of insects attacked. The vast majority of plant feeders fall into the gall wasp family Cynipidae, with relatively fewer species in the superfamily Chalcidoidea. There are but a handful of gall-forming braconids (Braconidae), and the only known plant-feeding ichneumonids (Ichneumonidae) feed on pollen provisions given to bees, and are thus actually cleptoparasitoids (Wharton and Hanson 2005).

Seed feeders

When speaking of seed-feeding Hymenoptera, we are not talking about wasps that devour seeds as do birds or live in the soft tissue surrounding a seed. We are referring to the larva of a wasp that resides inside the seed, or sometimes what appears to be a seed. As with all biological statements, preciseness of wording is important because it is nearly impossible to make a statement that is true without some sort of qualifying verbiage. As a simple example, a seed-feeding wasp would be one that lives inside a seed in a rose hip, but not in the rose hip tissue itself. The wasp is an internal seed feeder. This seems relatively straightforward, but the matter becomes more complicated because seeds occur in many different forms. Rose seeds are what they seem to be, that is, seeds. But the apparent seed of holly (*Ilex*) and sumac (*Rhus*), just to name two plants, is not actually the seed, but a nutlet composed of a hard covering called the endocarp with the seed inside of it. The combination is a pyrene, which is comparable to the seed-containing pit of a peach. Several common wasps in the United States feed externally on the seed of holly and sumac, but they are inside relative to the endocarp. Thus, technically they are external seed feeders, but in reality they would simply be called seed-feeding wasps by most.

The preceding belabored description was engineered not out of spite, but because it is necessary to establish certain factual limits. Within Hymenoptera, seed feeders occur rarely and sporadically, and then only in the parasitic wasp group. This eliminates seed-feeding harvester ants, for example, because they consume seeds, they don't live inside them. There is no single family of Hymenoptera that consists exclusively of seed-feeding species, but they are found in limited numbers in a few families.

Because the ovipositor of a parasitic wasp is designed to penetrate either a substrate containing a host insect and/or the host itself, these needle-shaped ovipositors are ideal for inserting an egg through the exterior wall of a developing seed. Once deposited inside, the egg hatches, the larva develops on the nutritious tissue and pupates inside the seed wall, and an adult eventually chews its way through the wall into the open to repeat the process. Generally the wasp's development is in complete synchrony with the host plant's reproductive cycle. If it is a yearly cycle from flower to seed, then the wasp has one generation per year; if multiple flowering periods within a year, then the wasp has multiple generations.

Let's take an actual example of a holly tree (*Ilex*) and its seed-feeding parasitic wasp of the genus *Megastigmus* (Torymidae). The holly blooms in the spring, its seeds develop through the summer inside of newly swelling berries (technically drupes inside each of which are four seeds each inside its own endocarp, but I'm sticking with berries because my head hurts), mature in autumn, and overwinter until the next spring, at which point the tree blooms again. In the spring, adult wasps emerge from seeds of the previous year's berries. The berries may still be hanging on the tree or may have fallen to the ground. Once mated, a female wasp is capable of laying many eggs in many seeds, but she lays only a single egg onto each fertilized and developing seed of the holly (she has to oviposit though the endocarp wall, which will later be the seed's outer cover). A newly hatched larva consumes the seed but not the cover, it overwinters as a larva, and then it emerges the next spring as an adult to repeat the process.

Being Hymenoptera, however, there are a couple of wonderful modifications of the seed-feeding story. Describing holly seed, again, the first modification is that not all wasps from a previous spring's batch will necessarily emerge a year later in the next spring. A large proportion will—perhaps 80 or 90 percent—but some will remain as larvae or prepupae for yet another year in the holly seed. Even then, some of those will not emerge, but may remain for a third year. The ability to exist without food for two or three years is remarkable, but it does insure that should wasps emerge at a time when no developing seeds are available, even though they might all die out, at least perhaps their kin will survive until another year when conditions might be more favorable. I don't know what the world's record is for parasitoid larval longevity, but I once kept seeds of a California buckthorn (*Rhamnus crocea*) for four years, during which period at least one adult wasp (Eurytomidae) emerged each year.

The second amazing aspect of the seed-feeding story is a bit harder to swallow, literally. It is common knowledge that animals eat fruits and berries that contain seeds. The tissue surrounding a seed is easily digested, but the seed itself, not so much so. Therefore, it is common for a large percentage of seeds to pass unharmed through the digestive systems of animals. Some even require the experience to germinate (that is, acid scarification). It should come as little surprise, then, that a wasp larva residing inside a seed might also experience the joys of a gastrointestinal roller coaster ride. Surprisingly, it will survive to tell the tale. In studies done on the rose seed chalcid (*Megastigmus aculeatus*, Torymidae) passed through bird intestines, there was no difference between emergence of wasps that rode the bird-gut trail and controls kept bird free.

Seed-feeding Hymenoptera are found principally in two families, Torymidae (Figure 73) and Eurytomidae, both in the superfamily Chalcidoidea, which contains more than 23,000 described species. Of these, fewer than 200 seed-infesting species are found among the torymids and eurytomids. A list of hosts of these two families includes fruit and nut trees (apple, pistachio, soursop, almond, apricot, plum, crabapple, guava, pandanus), shrubs (ceanothus, mountain ash, ninebark, rose, serviceberry, sumac), ornamental trees (Brazilian pepper, holly), vines (grape, Virginia creeper), herbs and vegetables (alfalfa, caraway, carrot, dill, fennel, parsley, parsnip), and evergreen trees (Douglas-fir, fir, juniper, hemlock, pine, spruce).

Galling parasitoids

A gall is a swelling caused by an insect, mite, nematode, bacterium, fungus, or virus feeding within a plant's tissue. Gall formation is caused by several different physiological processes acting alone or in combination and may include proliferation and/or enlargement of plant cells. In the case of wasps, galls may be induced either by larval feeding or by chemicals introduced when the female lays an egg. Generally the presence of a gall does little or no harm to a plant, and some even have an ornamental value in themselves, somewhat like small, colorful sculptures or even bangles hanging from a tree. Because a gall increases the volume of plant tissue, other insects, called inquilines, may feed internally near the gall former without harming it. Gall formers also present a nicely packaged bundle of protein that cannot flee when attacked. And they are attacked, endlessly, by hundreds of species of hymenopterous parasitoids as well as other animals such as birds. One final note about galls themselves is that after the creator abandons its home, other insects may move in if it is large enough. Solitary bees and wasps commonly use old gall cavities in which to build a nest, and some ants do so as well. Then these insects, themselves, are susceptible to being parasitized. Therefore, in galls there are many levels of occupancy involving many insects over time.

In the world of insects, gall formers may be found scattered among a few families and orders, but the majority are found in two preeminent groups, namely flies (Diptera, especially the family Cecidomyiidae) and Hymenoptera. Within Hymenoptera the great majority of gall formers fall into the parasitic wasp group, especially the family Cynipidae, a numerically large family containing nearly 2500 species worldwide. Gall formers may also be found among the sawflies (Tenthredinidae), but Cynipidae is the preeminent family of Hymenoptera whose entire caste of species, with few exceptions, is devoted to the pursuit of making plants swell up (Figure 74). We'll investigate a few additional families of parasitic wasps, but the gall wasp family must take the lead. Within the realm of gardening it is more likely that a gardener has seen the gall of a wasp than its creator. Truthfully, the appearance of the galls these wasps make are generally much more interesting and diverse than the wasps themselves.

Although you might reason that a plant swelling, by any other name, is just a plant swelling, you'd be far from correct. With the large number of gall wasps, a glimpse at the hundreds of different gall shapes, colors, and plant parts affected would allay any thought of simplicity when it comes to cynipid galls.

Where a female gall wasp lays an egg is entirely species specific. Some lay eggs in plant roots, others in stems, buds, leaves, flowers, or acorns. The galls thus formed may be integral to the plant and not separable (and sometimes not easily seen), or they may be an obvious outgrowth easily removed from the plant. A leaf gall, for instance, may appear to be a tiny blister in the leaf surface or it may appear to be a crown sitting upon the leaf. In the former instance it is called an integral gall because it cannot be separated from the plant, and in the latter case it is called a detachable gall because it can be easily removed. Gall size is specific to the wasp, with some species causing minute galls $1/16$ inch (1.5 mm) in diameter and some creating spheres up to 1 inch (25 mm). Then there is the California oak gall wasp (*Andricus californicus*), whose apple-sized galls may reach as much as 5 inches (125 mm) or more across (Figure 75). In general smaller galls are the result of a single larva living within the gall, called a monothalamous gall, whereas large galls such as the California oak apple are created by a cluster of larvae and are called polythalamous galls. The exact number of occupants is dictated by the female and is species specific. The size of the gall may not indicate its true nature, as some of the larger, spherical galls may be as solid and hard as a tree trunk, whereas others may be hollow, with a shell as brittle as an egg and a single wasp suspended in the center by delicate filaments of plant tissue.

As to shape and color, there are hundreds of possible combinations, many of which were given names by one of our foremost cynipid experts, Lewis H. Weld. His terminology for identification of galls included such descriptors as biscuit shaped, pill-box shaped, onion shaped, spindle shaped, two-story red cells, stony hard mass inside acorn, swollen filament, wrinkled bullet, round bullet, club shaped, globular with nipple and pink stripes, red-brown spiny, red spangle, flat spangle, woolly, crenulate, midrib cluster, blunt horns on rim, covered with short yellow fuzz, greenish with raised purple spots (I recall having these symptoms as a child), and a host of others. Some galls are named for the shape they represent, such as the acorn-plum gall (Figure 76), dried peach gall, and oak apple. There is one produced by a tiny leaf-galling wasp (*Neuroterus saltatorius*) in California that Weld called flea seeds, because the entire gall can jump, much as jumping beans, when disturbed. Another name I've seen for this species is bewitched mustard seed, which gives you some idea of size as well as the imaginative names given to some of these galls.

Some cynipids have a complicated and nearly unique system of alternating sexual generations between one of males and females (bisexual) and another of all females (asexual). The bisexual generation makes short-lived galls in the spring, generally in newly swelling buds and flowers and small in size. These galls are entirely different from the galls of the next generation, the asexual one, in which galls appear on different plant parts (as determined by the female) and are larger. Both the adults and the galls of these generations are so different that in many cases early researchers thought they were caused by different species and even genera of cynipids. There is no need to feel ashamed if none of this makes sense to you, because it is still baffling to many researchers to this day.

Staying with the cynipid wasps for a moment, we should mention their hosts plants. The vast majority of species are found in two plant families, Fagaceae (primarily oaks) and Rosaceae (roses), where their galls are usually conspicuous, often bizarre, and sometimes abundant. Among oak relatives, galls may also be found on tanbark oak (*Lithocarpus*) and chinquapin (*Castanopsis*). A few members of the Rosaceae are the next most common group of hosts, some of which may be found in the garden, including galls on stems and roots of rose, blackberry, raspberry, and potentilla. Roses also have leaf galls. Strawberries, both native and cultivated, are another rosaceous plant attacked by gall wasps, which induce a polythalamous petiole swelling of the leaves.

After oaks and rosaceous plants, the few remaining cynipid gall formers are mostly associated with Asteraceae, galling plants such as *Chrysothamnus, Hypochaeris, Lactuca* (wild forms), *Lygodesmia, Microseris, Pernanthes,* and *Silphium.* One common asteraceous plant certainly known to everyone is the dandelion, introduced into North America from Europe as was its cynipid gall wasp, *Phanacis taraxaci.* This wasp creates clusters of galls on a leaf's midrib. Another North American cynipid introduced from Europe, *Liposthenes glechomae,* causes leaf galls on ground ivy (*Glechoma hederaceae*), a common weed throughout much of the continent. I must confess that when I lived in Maryland, I grew both dandelions and ground ivy with the greatest skill known to a gardener, but I've never seen either wasp. Apparently I must have done something wrong.

Having discussed the most numerous and exclusive of the gall-forming parasitoid species, the gall wasps, it now remains to treat a few of the smaller, less well known, and less likely to be seen gall-inducing wasps. These families are not composed entirely of gall formers, or even plant feeders, but contain a mixture of feeding types from true parasitoids to inquilines, seed feeders, and gall makers.

The Tanaostigmatidae (Figure 77), with fewer than 100 species throughout the world, is second only to Cynipidae in its number of purported gall-forming species. Because of its small number of species and esoteric choice of host plants, this family remains virtually unknown to entomologist and gardener alike. In the United States, the family occurs in the southwestern deserts and southern Florida, where it causes stem, leaf, or seed galls on shrubby legumes including *Acacia, Mimosa, Pithecellobium,* and *Prosopis.* In other parts of the world some species in this family are thought to be true seed feeders, inquilines, and even parasitoids, but their life histories are far from confirmed.

Eurytomidae is a family with more than 1500 species worldwide. Of these, about 75 species in three genera (*Tetramesa, Cathilaria, Erytomocharis*) are gall formers in grass stems (Poaceae), including cereal grains such as rye and wheat as well as bamboos. One species of *Eurytoma* causes galls in pine stems, and another, the orchidfly (*Eurytoma orchidearum*; Figure 78), causes swelling in orchid roots. Its origin is likely South America, but it is known to occur in Hawaii and Florida. In other regions of the world eurytomids cause galls on well-known plants including stems and roots of several species of citrus and roots of *Philodendron* and *Dieffenbachia*.

Pteromalidae, a huge and complex family with more than 3500 species worldwide, has but a single known gall former, but one of importance. The blueberry wasp (*Hemadas nubilipennis*) causes galls on vegetative shoots of *Vaccinium*.

Eulophidae, another large and complex family with more than 5100 species worldwide, has no known native gall formers in North America, but several introduced species have started showing up in the United States since the mid 1990s. In Hawaii, an introduced African species is galling the leaves and stems of native *Erythrina* trees. In California, one introduced Australian species causes galls in seed capsules of introduced *Eucalyptus*, another causes blister galls on leaves of *Eucalyptus*, and yet a third causes galls on leaves and stems of commercially grown waxflower (*Chamelaucium*), itself introduced from Australia. In Australia there appear to be many such gall-forming eulophids just being discovered, and they are rapidly spreading around the world with Australian plants being brought into horticultural markets.

Although the ichneumonid (Ichneumonidae) and braconid (Braconidae) families contain more than 40,000 described species—and countless more to be discovered—less than a handful of braconids are known to be gall formers and these are found only in the American tropics and Australia (Wharton and Hanson 2005). Galls are formed in legume seeds and pods, melastome fruit, leaves and stems of banksias and proteas, and philodendron leaves. Although braconids have been reared from other galls, there is some evidence that they are actually attacking the true gall former and not causing the galls themselves (Wharton and Hanson 2005). No ichneumonids have yet been shown to cause galls, but several are thought to consume gall tissue created by other insects. I mention these families only because there are so many species, and we know so little about them, chances are some gall formers will eventually turn up in our gardens.

Fig wasps

I include fig wasps as a separate category within plant-feeding parasitoids because of the distinct curiosity most folks, including gardeners, have for them. The biology of these wasps almost ensures that they are neither fish nor fowl when it comes to being seed feeders or gall formers. They appear to be both! Fig wasps belong to the Agaonidae, a family currently surrounded in such a confusion of taxonomic, behavioral, and ecological contradictions it's nearly impossible to discuss in a rational manner. At present the family consists of almost any wasp in the superfamily Chalcidoidea that feeds within a fig (*Ficus*). Therefore it contains both plant feeders and the parasitoids that feed upon them. Most experts agree that the parasitic forms belong in other families—yet to be determined—and that Agaonidae should only contain the true fig-feeding species, currently all placed in the subfamily Agaoninae. More than 350 species of true fig-feeding wasps have been described. We need concern ourselves with only the latter category here, so when I say "fig wasp," I mean a plant-feeding fig wasp.

Female fig wasps of various species appear similar to each other, especially in having an elongate, flattened head with flattened mandibles, each equipped with an appendage with numerous teeth (Figure 79, left). Males, on the other hand, are the most curious of all wasps, being grotesquely modified and appearing more like a worm than a wasp (Figure 79, right). The males bear absolutely no resemblance to females of their own species. Their abdomens are elongate, telescoping, and curved under the body; their legs are greatly enlarged; they are incapable of flight, being wingless or having reduced wings; and their eyes are reduced to almost nothing. Males can't leave the fig in which they mature. Their only job is to chew a hole into a nearby flower within the fig containing a female, fertilize her, and die.

Figs and fig wasps share a mutualistic association. The former can only be pollinated by the latter, and fig wasps can only complete development in figs. According to one report, of all insects, fig wasps are "the insect group showing the greatest host specificity" (Schneider 1997). It is generally accepted that fig wasps are specific to fig species. As there are nearly 800 known fig species, it seems likely that there will be a similar number of pollinating fig wasps, although it is becoming evident that some fig wasps pollinate several different species of related figs.

In the continental United States, there are only a couple of native fig species, both occurring in southern Florida. The fig wasp *Pegoscapus jimenezi* pollinates the strangler fig (*Ficus aurea*) and *P. asseutus* pollinates the shortleaf fig (*F. citrifolia*). Several other introduced fig wasps are known from Hawaii, California, and Florida on introduced fig species. These include *Eupristina verticillata* in the laurel fig (*F. microcarpa*, Florida, Hawaii), *E. altissima* in the false banyan (*F. altissima*, California, Florida), *E. masoni* in the banyan tree (*F. benghalensis*, Florida), *Pleistondontes greenwoodi* in small leaved fig (*F. obliqua*, California) and Jamaican cherry (*F. americana*, California), *P. froggatti* on Moreton Bay fig (*F. macrophylla*, Hawaii), and *P. imperialis* on banyan (*F. platypoda*, Hawaii; Boucek 1997; Insects of Hawaii: http://www.hear.org/starr/hiinsects/index.html). The most famous of all American fig wasps, however, is not even American. *Blastophaga psenes* was introduced from Turkey to pollinate the edible commercial fig, *F. carica* (Smyrna or Calimyrna fig), itself introduced from Europe and now grown commercially extensively in California.

To understand what fig wasps actually do, we must first examine a fig, which is a complex structure that is both an accessory and multiple fruit called a syconium (Castner 2004). A syconium is a receptacle upon which is placed hundreds of flowers (imagine a sunflower) and then is folded into a hollow ball with a tiny opening at its apex called the ostiole. It's somewhat as if you implanted hundreds of tiny flowers in the palm of your hand and then made a fist. When you eat a fig you are eating a receptacle and hundreds of its flowers, which are the actual fruit of a fig—but more of that in a bit.

Over the years I have found that there is generally great confusion (mostly mine) surrounding fig development and fig wasps. This confusion arises because of two factors. First, there are hundreds of cultivated horticultural varieties of the commercially grown fig, *F. carica*, and, second, the figs of these *F. carica* varieties are placed into three classes that develop in different ways: with pollination by wasps (Smyrna class, or caducous figs), without pollination by wasps (persistent, common class, or parthenocarpic figs), and with an early unpollinated crop and a later pollinated one (intermediate, or San Pedro class). The direct and easy lesson is that the dozens of figs commonly available and cultivated by home gardeners need no pollinators (that is, they are parthenocarpic). These include such varieties as Brown Turkey, Conadria, Kadota, Mission, Celeste, Verte, Kadota, Panache, and White Genoa, just to mention a few. For home gardeners the pollination problem is solved.

Alternatively, the commercially cultivated, edible variety of fig that everyone refers to as the fig, famous for Fig Newtons, goes by at least two names. Originally the name given to the edible fig in North America was the Smyrna fig, based presumably on Smyrna, the city in Turkey from which the fig wasp *Blastophaga psenes* was first introduced. Because California produces nearly all of the commercial figs in the United States, the Smyrna fig eventually became the Calimyrna fig. This fig is the type that needs pollination, and it involves a complicated series of steps to produce a crop. It also requires a complicated set of terms to describe how pollination takes place. For those interested in such things, I present a simple overview. For those who don't, I suggest skipping to the next section.

Calimyrna figs (that is, the syconia) grow on two kinds of trees: a female tree producing edible figs (sometimes called seed figs) with only long-styled female flowers and a male tree producing inedible figs (caprifigs or gall figs) with pollen-bearing male flowers and short-styled female flowers (called gall flowers). I'll refer to these as edible and inedible figs to simplify things because things only become worse as we proceed. Having both male and female flowers, the inedible fig is constructed differently from the edible fig. Just below the fig's opening (ostiole), male flowers line a narrow channel until it opens into an interior chamber lined with female flowers. Herein live the fig wasps, both male and female. When the male wasps mature, being wingless, nearly blind, and probably deaf and dumb as well, they mate with females, which then emerge from the inedible fig, passing through the gauntlet of male flowers, out the ostiolar opening, and into the open air. They pick up pollen as they exit. At this point females are free to fly to and enter any fig they find, either edible or inedible.

The tricky bit is that fig wasps can only live and reproduce in the female short-styled flowers of the inedible figs, not in long-styled flowers in the edible figs. This is because the female fig wasp inserts her ovipositor down the style of a flower to lay an egg into its ovary. Her ovipositor can reach the target in a short-styled flower but not a long-styled one. Thus, even though she attempts to oviposit into flowers of both kinds of figs, either edible or inedible, her eggs will only reach and survive in the ovary of the inedible fig. In either case, the outcome is still death for the female because as she enters the fig of choice through its ostiole her wings break off, and sometimes other bits and pieces as well. She cannot exit the fig but she can still attempt to lay eggs, and even if unsuccessful, she still pollinates the fig she has entered.

Now, enter the fig farmer. Knowing all about the above wasp perambulations, the farmer segregates his trees into two areas: the edible fig production area and the inedible wasp production area. When the fig wasps are ready to emerge from the inedible fruits, the farmer picks and bags a few and places them into the edible fig production area, being careful not to overdo it. These females pollinate the edible figs, which then increase in size as their fertilized ovaries swell. The farmer must be careful not to introduce too many wasps or the figs will become overpollinated, balloon up, and split, thus ruining the crop. That is why the two types of fig trees are not interplanted. Too many wasps doing too good a job of pollination is too much of a good thing.

If you've followed this discussion so far—and I barely have—several still unanswered questions might have come to mind. The likely burning question is, "If fig wasps go into a fig but don't come out, am I eating wasps when I eat commercial figs." The answer is yes, but not so you'd notice! Any wasp bits that remain in a fig once she dies are broken down by the protein-digesting enzyme ficin, produced by the fig. The wasp's atoms and molecules are still there but not the wasps as you know them. When you hear the crunchy bits of a fig it is not a wasp, it is the achene, a hard thin wall surrounding each seed (the outer cover of a sunflower seed is its achene). Whether a wasp is present or not, neither fig seeds nor fig wasps crunch. Achenes do!

Another question that might come to mind has to do with my statement that fig wasps appear to be both seed feeders and gall formers. When the female fig wasp lays an egg down the style of a short-styled flower, she places it within the seed, which is eventually eaten by the wasp larva. Additionally, the ovary becomes swollen and is transformed into a gall. Thus a fig wasp is both a seed feeder and a gall former.

The details given above relate primarily to but a single species of pollinating wasp and fig tree. In truth it is highly oversimplified to as basic a level as possible, and it's still complex. It is believed that half of all fig species are of the Calimyrna type (that is, gynodioecious), and the other half are of the monoecious type without separate male and female trees. Some of these figs are known to have four layers of flowers each supporting different sets of plant-feeding and parasitic wasp species. These wasps, both plant feeders and parasitoids, total less than 800 described species, but they are so poorly known it is likely that thousands more remain to be discovered. When considering the hundreds of fig species and hundreds of wasp species, all of which have coevolved into hyper-complex relationships, one can only wonder at the intricate universe yet to be found in a simple fig.

Stinging Parasitoids

Compared to true parasitoids, stinging parasitoids are limited in numbers of species, biological behavior, and the range of hosts they attack. Stinging parasitoids are represented by 12 families (plus the spider wasp subfamily Ceropalinae) and about 13,600 species. These families fall within the predatory wasp group (Apocrita; see the table of hymenopteran families). Most are unknown to the gardener, but the velvet ants (Mutillidae), sometimes called cow killers, might be familiar to some. Although true parasitoids tend to be small, fragile, gracile, and incapable of stinging, many of the stinging parasitoids reach the opposite extreme: large, robust, clumsy, and quite capable of stinging. They are not called stinging parasitoids for nothing! Even some of the smallest can sting—mostly to no effect—and even males of some species have spines at the tip of their abdomen that allow them to perform a false-stinging routine. Unfortunately for the neat and tidy, there are stinging parasitoids that cannot sting (for example, cuckoo wasps, Chrysididae), which is certainly an oxymoron, but as we have seen many times before, Hymenoptera are not easily stereotyped.

The biological range of stinging parasitoids is limited, at least compared to true parasitoids, and so we'll examine this group from the host's point of view.

Bee and Predatory Wasp Larvae

Ironically, the largest group of hosts attacked by stinging parasitoids are their own stinging cousins the predatory wasps and bees. The groups responsible for this are the velvet ants (Mutillidae) with about 5000 species worldwide, members of the subfamily Chrysidinae (Chrysididae) with about 2200 known species in the subfamily, and the rarely encountered family Sapygidae with 80 species. The sapygid wasps are scarcely worth mentioning as they are few and far between and rarely seen by even the most dedicated entomologist. They are noteworthy for but a single reason. Females lay their eggs in nests of solitary bees and the resultant larva consumes both the bee larva and its pollen provisions. Pollen-feeding parasitoids are basically unheard of, though a small group of ichneumonids are known to do so.

The most likely stinging parasitoid known to the general public would be female velvet ants, which are wingless and commonly seen wandering over the ground, appearing much like ants (Figure 80). They derive their name from the abundant hairs covering their body, which in most cases gives the impression that they are covered in velvet. Some species, such as the thistledown velvet ant (*Dasymutilla gloriosa*; Figure 81), are less well groomed in appearance. Female mutillid ants have exceptionally long stingers and inflict horribly painful stings. Based on my experience, I don't know how any other hymenopteran could hurt worse even if pain is a subjective matter (spider wasps, Pompilidae, are certainly a contender). Male mutillids are winged and generally appear nothing at all like the female. Although the biology of most species is unknown, the hosts commonly associated with these wasps are mature larvae, prepupae, and pupae of ground-nesting solitary bees and solitary predatory wasps. In cases where the host has spun a cocoon, the mutillid bites a hole in it, lays an egg within, then seals it back up. Although bees and wasps are the primary hosts of velvet ants, a few species attack other insects; perhaps most interestingly in Africa several species attack pupae of the tsetse fly.

One of the most commonly seen velvet ants in the eastern United States is the cow killer (*Dasymutilla occidentalis*). The host of this large (up to 1 inch, 25 mm) black-and-red wasp is somewhat of a mystery as it is claimed to have been reared from bumble bee larvae and pupae (*Bombus*, Apidae), cicada killer larvae (*Sphecius speciosus*, Crabronidae), as well as pupae of the horse guard (*Stictia carolina*, Crabronidae; Eaton 2005). Such a large, common, spectacular parasitoid should have little question concerning its hosts, but such is our generally poor knowledge of parasitoids that glaring questions still remain.

At the opposite extreme from velvet ants are cuckoo wasps (Chrysididae; Figure 82), which are hairless, small, metallic green or blue, highly sculptured parasitoids sometimes seen scurrying about exploring for nests in crevices, walls, or plant stems. These shining jewels among the insect world have two characters unique to their kind. The first will likely prove confounding—these are nonstinging, stinging parasitoids. That is, their ovipositor has (d)evolved into a prehensile tube through which the egg is laid. The second character of interest is that cuckoo wasps curl up into tight balls (much like pillbugs) when attacked. Because they enter directly into the dangerous nests of predatory wasps and bees, but have no stinger themselves, their only means of self preservation is by being armor plated and curling up into a protective ball.

Although called cuckoo wasps, evidence suggests that most are actually true parasitoids that attack the host larva directly and not cleptoparasitoids, as their common name suggests. Unlike their cousins the mutillids, chrysidid wasps are less attracted to ground-nesting bees or wasps, but instead search out those that nest in stems and twigs, mud cells, crevices, holes in walls, and even old galls. The wasp enters a nest, lays an egg on the host larva, and departs. The chrysidid larva waits until its host has finished eating its supply of prey (if a wasp) or pollen (if a bee), allows the host to spin a cocoon—with itself still upon the host—then consumes it.

Beetle Larvae

Beetle larvae are the next most numerous group of insects attacked by stinging parasitoids. The two largest groups that specialize in these larvae are scoliids (Scoliidae) and tiphiids (Tiphiidae) with about 1800 worldwide species between them. These brightly colored black-and-yellow or black-and-orange wasps can reach up to 1.5 inches (40 mm) in length. Species of both families are external larval parasitoids of ground-inhabiting scarab beetle larvae (often called grubs), which are known to gardeners as lawn pests as immatures and as plant-feeding pests (for example, Japanese beetles) as adults. A tiphiid female (Figure 83) burrows into the soil, stings and paralyzes its host, lays an egg on it, and departs. Scoliids (Figure 84) do the same, but they build a cell around the larva, thus approaching their cousins the predatory wasps in constructing a sort of nest-like environment in which its larva is protected. In part, host stinging and a tendency to physically hide a host is what places the stinging parasitoids in a group somewhere between parasitoids and predatory wasps.

Some tiphiids (Methochinae) attack tiger beetle larvae (Cicindelidae) in much the same manner as does the chalcidid wasp (*Lasiochalcidia*) that attacks antlions. A tiger beetle larva lives in a burrow in the ground and attacks insects that wander near its opening. The tiphiid wasp walks up to the tiger beetle larva, allows the larva to grasp it in its jaws, and then stings it under the head. Once paralyzed, the wasps lays an egg on the larva, then backfills the burrow with soil.

Beetle larvae are also attacked by some members of the family Bethylidae, some of which are less than 1/8 inch (3 mm) in length. These larvae are most often associated with the stored grain and legume products found in our pantries. If used infrequently—or kept in poorly sealed containers—these products may become infested by grain beetles, flour beetles, drugstore beetles, and cigarette beetles. In turn, these pests may be attacked by stinging (as well as true) parasitoids. Some bethylids are known to move their host to a sheltered position and even guard their larvae until they reach maturity.

Bethylids, despite their small size, are capable of inflicting painful stings and even swelling in susceptible victims. If foodstuffs are improperly stored or housekeeping is left to ruin, large populations of beetles can build up, in which case bethylids will also increase. When this happens they, too, can become a nuisance, and it is possible to encounter multiple stings without actually seeing the culprit or realizing what it is.

Leafhoppers, Treehoppers, and Planthoppers (Hemiptera)

Next to the large numbers of beetles attacked, the most numerous host group is the true bugs (Hemiptera), including leafhoppers (Cicadellidae), treehoppers (Membracidae), and planthoppers (Issidae, Fulgoridae, Achilidae, Flatidae). All of these are attacked by two families, Dryinidae, with about 1100 species worldwide, and Embolemidae, with only 10 species worldwide. All these wasps are essentially unique in that their larvae are internal feeders within the host, but at some point the successively shed skins of the larva break through the abdominal wall of the host (much like a hernia) and form what appears to be a hardened, cyst-like sac (as "parasitoid sac" in the table of hymenopteran families). Some hosts are attacked in the nymphal stage and others in the adult stage. The host may actually survive to adulthood and be capable of limited reproduction, but generally their sexual organs degenerate to the point of uselessness. When the parasitoid is mature, it emerges from the host, crawls away, and pupates in a protected spot. Another unique aspect of many dryinid wasp females is that their claw and last leg segment (tarsus) are modified to serve as pinchers. These pinchers are used to grasp a host nymph or adult as the female inserts an egg between the abdominal segments of its victim.

Spiders

Essentially all members of the spider wasp family Pompilidae are spider killers, but they do so in different ways. The majority of pompilids (about 4000 species worldwide) are true predatory wasps, but members of the single subfamily Ceropalinae (200 species) act as stinging parasitoids. A few of the other 4000 species act as parasitoids as well, but the Ceropalinae are believed to be parasitic in total. It is confounding groups such as pompilids that make discussing Hymenoptera such a challenge.

Members of the Ceropalinae adapt different methods to achieve their goals as stinging parasitoids. In its simplest form, a female simply lays an egg on the host spider—after temporarily paralyzing it—and the spider then wakes up and is eaten alive as it attempts, unsuccessfully, to conduct its daily business. In other cases, the ceropaline locates a true predatory pompilid that is dragging a spider prey to its nest. The camp follower lays an egg on the spider, and its predatory cousin, none the wiser, drags the spider and uninvited guest into its nest. The parasitoid egg hatches first, the larva destroys the host egg or larva, and then dines on its spider supper.

Miscellaneous Hosts

Some of the stinging parasitoid families contain subfamilies that differ with respect to the main host choices of the family. One example is the subfamily Amiseginae (Chrysididae) in which all known species attack the eggs of walking sticks (Phasmatoidea). This represents the only group of stinging parasitoids that attack eggs. Another specialist subfamily of Chrysididae is Cleptinae, which attacks the prepupae of sawflies by biting a hole in a cocoon, inserting an egg, then sealing the cocoon up.

Within the family Bethylidae, the subfamily Bethylinae specializes in moth larvae. Much as with beetle larvae, moth larvae such as the Indian meal moth (*Plodia interpunctella*) and flour and grain moths are pests of stored products. These larvae are attacked by bethylid wasps and sometimes may be found in the home. Additional hosts include larvae of clothes moths and plant-feeding moths, such as leaf tiers, leaf miners, leaf rollers, case bearers, and shoot and fruit borers.

Members of the Rhopalosomatidae are all thought to attack immature crickets (Orthoptera) with their larvae attached externally to the abdomen. Very little is actually known about the biology of this family, though it is not especially uncommon. Other poorly known families include Sclerogibbidae that attack webspinners (Embiidina), seldom seen insects that live in silken galleries under rocks, in leaf litter, or under bark. Although relatively common, species of Brady-nobaenidae have only one known host record. They have been found externally on immature solpugids (Arachnida, Solifugae). The family Sierolomorphidae has never had a host record, and perhaps with a name like that, it never will.

Pitiless Indifference

The parasitoid life cycle, being gruesome in the extreme, was actually an impetus for one of our most distinguished biologists to consider what has seemingly become one of the greatest debates of our time. Charles Darwin (1860), in a letter to Asa Gray, the most important American botanist of the nineteenth century, wrote:

> There seems to me too much misery in the world. I cannot per-suade myself that a beneficent & omnipotent God would have designedly created the Ichneumonidae with the express inten-tion of their feeding within the living bodies of caterpillars, or that a cat should play with mice. Not believing this, I see no ne-cessity in the belief that the eye was expressly designed. On the other hand I cannot anyhow be contented to view this wonder-ful universe & especially the nature of man, & to conclude that everything is the result of brute force. I am inclined to look at everything as resulting from designed laws, with the details, whether good or bad, left to the working out of what we may call chance. Not that this notion at all satisfies me. I feel most deeply that the whole subject is too profound for the human in-tellect. A dog might as well speculate on the mind of Newton.— Let each man hope & believe what he can.

Although Charles Darwin has received much contentious press of late, any right-thinking person reading that statement could argue that Darwin was not such a bad fellow after all.

A more recent comment on the evils of parasitoids was provided by Richard Dawkins in his book *River Out of Eden* (1995), who put it more abstractly when discussing acts such as those specifically at-tributed to Ichneumonidae. He pointed out that tens of thousands of parasitoids and predatory wasp species exist throughout nature, ensuring that other organisms suffer the action of being eaten alive. "This sounds savagely cruel," he wrote, but "nature is not cruel, only pitilessly indifferent. This is one of the hardest lessons for humans to learn." Parasitic wasps, it would seem, are far more than life forms simply once dismissed as "the green myriads in the peopled grass" (Pope 1734). They are part of our existential psyche. Who would have thought it?

7

The Garden's Wolves:
Predatory Wasps

SOLITARY PREDATORY WASPS are also sometimes called hunting wasps, because they actively hunt and capture prey for their young; digger wasps, because many burrow into the ground to build their nests; or sand wasps (Figure 85), because many prefer to nest in sandy soil. Eusocial predatory wasps, such as yellow jackets, hornets, and paper wasps are often called by names as well, but they are unfit to print in books such as this. Gardeners are mostly unaware of the solitary wasps, but they are certainly familiar with these latter wasps, mostly for negative reasons associated with pain. From personal experience I can sympathize with this view, but it really is unfortunate that one relatively small group of wasps should taint not only their own good work as predators, but also the work of thousands of solitary wasp species around the world.

There are over 17,000 described species of predatory wasps living throughout the world. This is roughly one-fourth as many species as the parasitoid wasps, by the way, but when the two groups are combined, more than two-thirds (106,000 species) of all Hymenoptera fall into the category of predators or parasitoids of other insects. And that doesn't even include ants, which are difficult to categorize. The resultant interactions of these carnivorous insects are a large part of what keeps insect populations in balance.

Among the champions of predatory wasps, Howard Evans and Richard Bohart produced some of the most comprehensive literature of our time. Evans wrote not only scientific papers but was foremost in popularizing wasps, having written *Wasp Farm* (1964). *The Man Who Loved Wasps* (2005) was a later compilation of his writings published posthumously by his wife, Mary Alice Evans. I highly recommend both (as well as *A Naturalist's Years in the Rocky Mountains*; Evans 2001). Throughout the discussion below on solitary wasps, especially dealing with Crabronidae, I cite information from one of Evans less popular titles (Evans 1966) and the comprehensive work *Sphecid Wasps of the World* by Bohart and Menke (1976).

In spite of Evans' devotion to the subject, he wrote of predatory wasps in *Wasp Farm* that "none of them make a very important impression upon man or upon the populations of insects they prey upon." I agree wholeheartedly with the first half of the statement, but with all due respect, I would argue that the second half is overly dismissive of the importance of predatory wasps in the balance of nature. Even the impact of a single group of predators such as yellow jackets or hornets is nearly unfathomable, but taken as a whole we can scarcely grasp the biological services provided by all the predatory wasps. Evans did make one very positive statement about predatory wasps, however, in that at some point they gave rise to the bees and ants, both groups of which are extremely significant to the natural world and to humans. So disagreements aside, one way or another predatory wasps and their descendants are hugely important in the scheme of things.

Although most gardeners are unaware of the lifestyles of even common wasps such as yellow jackets, they are woefully ignorant when it comes to solitary members of the assemblage. In fact, the life histories of lions, tigers, and bears might actually be more familiar, when none are to be found in our gardens, but predatory wasps certainly are. Unlike bees, whose larvae all gain their nourishment from the pollen and nectar of flowers, predatory wasps are extremely diverse in the variety of insects they feed their young as well as other aspects of their biology. These aspects can often help define the various families or even genera of wasps involved. Such attributes as nest substrate and architecture, type of prey used, method of prey collection, prey transportation, and number of prey items per larva all tell us something about different predatory wasps. If, for example, you see a wasp pulling a comatose spider across the ground, it is certainly a true spider wasp, which is the common name for the Pompilidae, all members of which use spiders as food for their immature stages. Alternatively, if you see a wasp stuffing a spider into a mud nest under the eaves of your house, it is a mud dauber of the family Sphecidae, not a true spider wasp. What a wasp is hunting and where it lives tells us a lot about what it is. It should come as little surprise that hunting and nest building by predatory wasps is the exclusive domain of the females. Most hymenoptera males are basically useless except for reproduction, and even then females can always create males without being mated.

Because of specific associations between predatory wasp and prey, this chapter is arranged differently from the chapters on sawflies and parasitoids. The solitary wasps, which can be defined largely by the type of host they hunt and method of nesting, will be grouped by hosts. The eusocial wasps, which are generalist predators and not defined by what they prey upon, will be examined separately. The family Crabronidae contains the most diverse assemblage of hunters and prey, consisting most often of adult insects. Other families such as Sphecidae and Ampulicidae are more restricted in their prey choice, and Pompilidae hunts only spiders. These families consist of solitary hunters. The family Vespidae is just plain chaotic with solitary predators, including the small, highly specialized subfamilies Masarinae (pollen provisioners) and Euparagiinae (weevil larvae provisioners), the very large subfamily Eumeninae (caterpillar hunters), and the eusocial hunter subfamily Vespinae (all sorts of prey).

Three final, but very important points need to be made about the predators we will examine in this chapter. First, the solitary hunters are highly specific in what they hunt, whereas the social hunters are not. Second, solitary hunters are generally mass provisioners in that they fill a cell with prey, lay an egg, seal it, and fly away. (A few feed their larvae somewhat progressively, though nowhere nearly as completely as do the eusocial wasps.) The larva is totally on its own, and the cell could as well be the larva's tomb if things go wrong. Social hunters are progressive provisioners that lay an egg in an open cell and feed the larva as it grows. In essence they are like birds caring for their young. Third, nests of solitary hunters are not to be feared. Rarely is the builder at home or even observed; only larvae and prey reside in the nest. In large nesting aggregations, many wasps may be flitting about but they have no interest in you, except perhaps to bluff their way past your large presence. Nests of social hunters, on the other hand, are continuously occupied and the occupants do not want to be disturbed. They are caring for their nurseries and, like any mother, will protect their young with their lives. There are varying degrees of aggressive protection, but the basic rule here is that it is not wise to stick your finger (or your face) into places it doesn't belong.

Predation, an Inconvenient Truth

To place the behavioral biology of predatory wasps in context, we must first admit to some failures, or perhaps subtleties, of definition. When discussing the true parasitoids we discovered that some larvae act more as predators than parasitoids. For example, a larva might feed by moving from host egg to host egg within a spider egg sack or from host mite to host mite within a gall, consuming prey as it moves about. This is not common, however, and so it is conveniently ignored when defining parasitoids and their behavior. Additionally, true parasitoids possess a flexible ovipositor down which an egg passes once the host has been subdued. This is the morphological definition of a true parasitoid preferred by experts. In all cases the female parasitoid simply finds a host, lays an egg (or eggs) on, in, or near it, and flies away.

Simplistically stated, solitary predatory wasps are generally defined by the fact that they actively collect one or more insects, carry the prey to a constructed place (a nest), lay an egg on the prey, seal the nest, and move on. Eusocial wasps differ in that they chew up their victims—as a decent predator ought to—then regurgitate it progressively to their larvae when they reach the communal nest. All predatory wasps are defined morphologically by the ovipositor, which serves only as stun gun/venom-injection apparatus. The egg is deposited from a genital opening at the ovipositor's base.

Thus, the difference between true parasitoids and predatory wasps should be simplicity itself, except that the stinging parasitoids display the biological habits of true parasitoids but have the morphological construction of predators. Thus, stinging parasitoids are inconveniently intermediate between the two groups and inconveniently difficult to place. Based on their biology I discussed them in the previous chapter, but based on morphology stinging parasitoids are members of the predatory group.

Solitary Hunters and Their Prey

Prey from giant cicadas to minute thrips are hunted by different wasp species, most of which have distinct and ritualistic behavioral patterns when it comes to species captured, methods of capture, transport, and concealment (that is, nest construction). Some predatory species spend less than a day to complete a single nest with a single cell, whereas others may spend weeks digging, constructing, and provisioning a single nest. The total amount of prey mass given to each wasp larva must be approximately equal to the final size the wasp will attain. This is achieved in different wasp species by providing one giant prey item or a number of smaller ones. Just because a wasp is large does not mean it must capture large prey. It probably does, but it can achieve the same objective by making more trips to gather up an equal amount of smaller prey. For example, in spider wasps (Pompilidae), the gigantic tarantula killer need provide only one spider for each for its offspring, whereas an aphid predator, for example, will need to gather many more.

Larval Predators

Among solitary wasps, two subfamilies in two families are particularly fond of insect larvae as food for their young: Ammophilinae (Sphecidae) and Eumeninae (Vespidae). The primary focus of these hunters is caterpillars of moths and butterflies, but beetle and sawfly larvae are also taken depending upon the species of hunter. Both groups, representing almost 3000 species worldwide, use larvae exclusively and are thus of great importance in keeping plant feeders under control (Figure 86).

Species of the common caterpillar-hunting genus *Ammophila* (Sphecidae) are found throughout the United States. They are thread-waisted, elongate wasps ranging in size from about 3/8 to 1 1/2 inches (10 to 40 mm) and colored from black and red (Figure 87) to solid orange (Figure 88). A similar genus with a single species (*Eremnophila aureonotata*) in the United States is solid black and as long as the longest *Ammophila*. Despite their imposing size, these wasps cannot sting.

An *Ammophila* wasp begins a nest by scraping at the soil surface with her mandibles or forelegs, depending in part on the species (Figure 29). Diving head-first into her work, she eventually entraps soil particles between the head and neck, or in the case of larger stones between the mandibles, backs out of the nest, and either walks forward and dumps the material a short distance from the entrance or flies in small, circular patterns, dumping as she flies. On a smooth surface, such as a stone terrace, it's possible to spot randomly dispersed soil and small pebbles accumulating a foot or so from the nest entrance. When the wasp has burrowed obliquely for 2 to 3 inches (5 to 8 cm) and prepared a terminal elliptical cell in which to place her prey, she does a remarkable thing. At least it seems remarkable to me. She locates a tiny stone about the size of the opening to her nest, picks it up in her mandibles, and inserts it. Then she proceeds to backfill the recently constructed burrow until no clue to its presence is visible.

After making a few orientation flights, she flies off, leaving a completely invisible nest inside of which there is nothing. This might seem odd, but in the world of predatory wasps it's just about normal. She has flown off to gather one or more caterpillars (depending upon her species), which she will paralyze but not kill by stinging certain nerve centers in the unfortunate victim's body. Now the wasp must get this prey back to her nest, which she does not by flying but by straddling the larva and dragging it to her nest, which may be hundreds of feet away. One instance of more than 65 yards (20 m) was described by Crompton (1955) in which the wasp had to drag her prey around hedges, down a street, then a pathway, under a gate, then through flowers and across a lawn to get home again. As Crompton noted, the wasp had not walked this route before but apparently only flown over it—a remarkable feat of navigation indeed, to be followed by an equally remarkable one.

The wasp must now find the completely camouflaged nest she had made before she went hunting. The larva is deposited near the nest entrance, and in a minimal amount of time the nest is found, the soil is scraped away, and the rock removed from the entrance. The wasp turns herself around, grabs the prey, backs into her burrow, deposits the prey in the terminal cell, lays an egg on it precisely where she wants it, backs out, and once again places a rock into the burrow entrance. She then neatly places additional particles over the rock and erases all signs of her presence by leveling the surrounding soil so that no evidence of the nest is visible. All of this is done in a matter of several minutes. Another remarkable thing is that some species of *Ammophila* have been seen picking up an additional rock, which they use to pound the soil firmly into place upon completing the nest, a rare instance of tool use by an animal other than humans. *Ammophila* are known to "maintain several active nests at once, yet remember the location of each and 'forget' each nest as it is completed" (Evans and West Eberhard 1970). Among predatory wasps, this ability is known so far only for this genus.

Eumeninae is the other group of caterpillar hunters. Beetle larvae are also used to a much less extent, and some records are known for sawfly larvae, but these are rare. The subfamily belongs to the family that contains the eusocial species (Vespidae) and at times has been placed in its own family (Eumenidae). In North America north of Mexico, there are some 250 species of eumenines. The majority are small, 1/2 inch (12 mm), and although they mostly go about their business unobserved, they are quite colorful with combinations of black, yellow, red, orange, or white. One common black-and-white species, *Monobia quadridens* (Figure 89), is fairly large and bulky for the subfamily, reaching nearly 1 inch (25 mm) in length. It could easily be mistaken for a bald-faced hornet or even a scoliid wasp. *Monobia* wasps nest in holes in wood, abandoned carpenter bee nests, and even empty mud-dauber nests. They provision with moth caterpillars and divide their cells with mud partitions. It is this latter habit that provides their common name of mason wasp.

Perhaps the most interesting euminines are the potter wasps (*Eumenes*; Figure 90), so named because they build small, sealed, jug-like pots from mud. These pots are placed on twigs, rocks, and even window screens. Inside each pot is placed several caterpillar prey hunted by the adult. Unlike many wasps that simply place their egg on the prey or at the bottom of the cell, potter wasps suspend their egg by a thread from the top of the pot. When it hatches, the wasp larva is hanging directly over its supper, and it remains attached to its safety line until the first caterpillar or two is consumed, then it is bold enough to drop down and feast among its hosts. Apparently the reason for this odd behavior is that female potter wasps only partially paralyze their hosts, which are still capable of some movement. A tiny wasp larva might be crushed if it had no way to retreat from its twitching dinner plate, and so it essentially becomes a trapeze artist.

Most eumenine wasps are not potters but usurp cracks, crevices, holes made by other insects such as wood-boring beetles, and even abandoned mud nests made by other wasps. Some species also dig their own nests or bore into plant stems. In these hidden places, the female wasp stores her prey, often partitioning the burrow into a number of cells, each containing prey larvae and a single egg. Some species make a nest with only a single cell. I once studied a eumenine species that simply dug a short vertical burrow in the soil with a cell at the bottom. The female would fly out, collect a caterpillar, fly back, drop the prey down the entrance—barely stopping to take aim—and then fly off looking for more larvae. As it turned out, the wasp was going to asteraceous flowers, digging into the flower head to find a well-hidden moth larva, extracting it from the flower, and returning home with the prey.

Those euminines that attack beetle larvae are much less common, but they prey on important families, primarily leaf-feeding beetle larvae (Chrysomelidae) and weevil larvae (Curculionidae). The tiny and rare subfamily Euparagiinae (10 species), also in the family Vespidae, is an exclusive hunter of weevil larvae. Female *Euparagia* excavate nests 3 to 4 inches (75 to 100 mm) into the soil and place 20 to 30 weevil larvae per cell. Unlike most ground nesters, females build periscope-shaped mud turrets over the nest entrance that help guide the wasp when she returns from the hunt. Females were observed working on the same nest 21 days after construction began and had completed four cells during that time. I mention such unsung rare wasps, in part, because these small creatures perform an amazing service in collecting large numbers of weevil larvae that would otherwise be damaging plants. I also mention it because it was the subject of my first published paper, coauthored with my classmate and long-time friend, Steve Clement.

Adult Butterfly and Moth Hunters

As commonly hunted as lepidopterous larvae are, very few solitary predators attack adult butterflies and moths. These rarely seen wasps are confined to about a dozen species of the genus *Stictiella* (Crabronidae), all residents of North America. Prey include nymphalid and lycaenid butterflies, skippers, and small moths in several families. The number of prey per cell varies from species to species, but 20 seems to be about the maximum. Nests are fairly deep, extending up to 11 inches (30 cm), and have only several cells. Adult prey are stored with the wings intact and are generally precisely placed upside down in a shingle-like arrangement so that the wasp larva can feed unencumbered as it proceeds down the carcasses without encountering appendages. When the wasp pupates, it is enshrouded in a sarcophagus of wings.

Adult Beetle Hunters

Possibly the largest group of adult beetle hunters in the world is the genus *Cerceris* (Crabronidae; Figure 91), with more than 800 species worldwide. The majority of species hunt beetles, but a few go after bees and other Hymenoptera. North American species hunt weevils (Curculionidae), flat-headed wood borers (Buprestidae), darkling beetles (Tenebrionidae), bean weevils (Bruchidae), and leaf-feeding beetles (Chrysomelidae). Each species of *Cerceris* is relatively specific as to which beetle type it takes. These wasps nest in the ground, either in flat areas or vertical banks composed of hard-packed sand or clay. Depending upon the species, nests range in depth from 1 inch (25 mm) to just over 3 feet (1.3 m). Most *Cerceris* species have the odd habit of collecting beetles and storing them in the burrow before digging the actual cell in which the beetles will be placed, whereas many wasps of other genera tend to construct a cell, then collect the prey and place it in the cell as collected. There have been reports of 40 to 60 beetles being placed in a single cell upon which the *Cerceris* larva will eventually feed. Generally somewhat fewer than 10 cells are made per nest. A related genus, *Eucerceris* (Crabronidae), also hunts beetles, and most of its nearly 40 species live in the American Southwest and Mexico. Its nesting habits are the same as *Cerceris*.

Adult Bee and Wasp Hunters

Philanthus species (Crabronidae; Figure 92) are sometimes referred to as bee wolves because most are hunters of adult bees. In Europe, one species (*Philanthus triangulum*) specializes in attacking honey bees, but in North America honey bees are exceptional prey for only a couple of the larger species. Most North American species prefer sweat bees (Halictidae), but it seems as if almost any adult hymenopteran will do in a pinch. Nests have been found to contain eumenid wasps (Vespidae), sphecid wasps (Sphecidae), cuckoo wasps (Chrysididae), ichneumonid and braconid wasps (Ichneumonoidea), and many families of bees. One nest of the California species *Philanthus zebratus* contained 25 different species of wasps and 20 species of bees in 17 brood cells. Frequently these prey are found nesting in close proximity, so perhaps this *Philanthus* was just lazy when it comes to hunting.

In *Philanthus* nests, after several prey are collected and placed in an underground cell, the last item has an egg laid on it. The cell is sealed off from the main burrow as other cells are dug and provisioned. Generally fewer than 10 cells are excavated per nest. Each nest may take several days to construct, and each wasp may construct several nests in a season. Some species, however, may only construct a single nest during an entire season.

Grasshopper, Katydid, Cricket, and Mantid Hunters

Many wasp genera provision their young with crickets, grasshoppers, katydids, and related groups, and they all belong to the families Crabronidae and Sphecidae. Species of two genera of predatory wasps (*Tachytes* and *Tachysphex*, Crabronidae; Figure 93) are predators of grasshoppers (Acrididae), katydids (Tettigoniidae), pygmy mole crickets (Tridactylidae), pygmy grasshoppers (Tetrigidae), and sometimes praying mantids (Mantidae). Apparently grasshoppers are one of the more common hosts. These genera are ground nesters, often using preexisting burrows abandoned by other wasps or even ants. Some nests extend nearly 3 feet (1 m) into the ground and have multiple cells. In some cases where a host lives in the ground (for example, pygmy mole crickets), the wasp will dig into the host's burrow and use its mandibles to grab the cricket by the head.

Figure 46. Some common sawfly (Tenthredinidae) adults can be quite colorful. The larvae of this *Tenthredo* species feed on a wide range of plants, but adults feed on small, soft-bodied insects and are good pollinators as well. Photo by Carll Goodpasture

Figure 47. Adult bristly roseslug (*Cladius difformis*, Tenthredinidae). Its larvae feed on rose leaves and receive their name for being somewhat slug-like and hairy. Length 3/8 inch (9 mm).

Figure 48. Adult female of a pine flower sawfly (*Xyela minor*, Xyelidae). Larvae of xyelids are primarily feeders in the staminate flowers of pine or in fir shoots. Length 1/4 inch (6 mm).

Figure 49. Adult melaleuca sawflies (*Lophyrotoma zonalis*, Pergidae) are being studied for introduction from Australia into Florida as a possible biological control agent against Australian paper bark tree, *Melaleuca quinquenervia*. Photo by Jason D. Stanley, U.S. Department of Agriculture, Agricultural Research Service, Bugwood.org

Figure 50. Larvae of a leaf-feeding sawfly (*Heteroperreyia hubrichi*, Pergidae) are being studied for introduction from South America as a biological control agent to combat the invasive Brazilian peppertree, *Schinus terebinthifolius* in Florida. Photo by Stephen D. Hight, U.S. Department of Agriculture, Agricultural Research Service, Bugwood.org

Figure 52. Adult of a purslane argid (*Schizocerella*, Argidae), larvae of which feed on *Portulaca* species. Most species are external leaf feeders, but some mine its leaves. Length 1/4 inch (6 mm).

Figure 51. Adults of the hibiscus sawfly (*Atomacera decepta*, Argidae) pose no threat, but their larvae cluster together when initially feeding on hibiscus. Unless controlled, the hibiscus becomes a skeleton of its former self. Photo by Bob Carlson

Figure 53. Larvae of elm argid sawflies (*Arge scapularis*, Argidae) are here engaged in synchronous defensive posturing to ward off enemies. Photo by Lacy L. Hyche, Auburn University, Bugwood.org

Figure 54. A mating pair of spruce webspinning sawflies (*Cephalcia arvensis*, Pamphiliidae), whose larvae feed exclusively on Norway spruce (*Picea abies*) in Europe. These sawflies are generally of little importance, but severe outbreaks occur during extended periods of warming. This species is not yet found in North America, but could easily become yet another pest of conifers if it does so. Photo by Andrea Battisti, Università di Padova, Bugwood.org

184 •

Figure 55. Adult male of the blackberry sawfly (*Onycholyda luteicornis*, Pamphiliidae). Larvae of this wasp feed in leaf folds of *Rubus* species. Length ³⁄₈ inch (9 mm).

Figure 57. Female of the introduced pine sawfly (*Diprion similis*, Diprionidae), accidentally introduced to North America from Europe in the 1910s. They have proven highly destructive to coniferous forests. Note the slightly saw-shaped antennae, indicating a female. Male antennae are much more feathery. Photo by John H. Ghent, U.S. Department of Agriculture Forest Service, Bugwood.org

Figure 56. A gregarious cluster of redheaded pine sawfly larvae (*Neodiprion lecontei*, Diprionidae). Many sawfly larvae are gregarious during their early feeding period. Photo by Lacy L. Hyche, Auburn University, Bugwood.org

Figure 58. Adult cimbicids (*Abia*, Cimbicidae) can be larger than bumble bees. The larvae of most feed on foliage of deciduous trees, but species of *Abia* feed on honeysuckle (*Lonicera*). Photo by Carll Goodpasture

Figure 59. The raspberry horntail (*Hartigia cressoni*, Cephidae) lays eggs at the tips of raspberry, blackberry, loganberry, and rose canes. Upon hatching, a larva girdles the tip, causing it to wilt, then burrows downward into the stem. Length ³/₄ inch (18 mm).

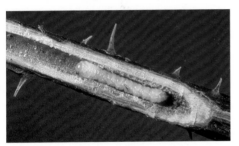

Figure 60. Larva of a rose stem sawfly (*Hartigia trimaculata*, Cephidae). Photo by James Solomon, U.S. Department of Agriculture Forest Service, Bugwood.org

Figure 61. Adult of the currant stem girdler (*Janus integer*, Cephidae), larvae of which bore into stems of currant (*Ribes*). Length ¹/₂ inch (12 mm).

Figure 62. An adult female horntail, also called the pigeon tremex (*Tremex columba*, Siricidae). Larvae of this wasp live under the bark of dead and dying hardwood trees. Despite its appearance, these wasps cannot sting. Photo by Steven Katovich, U.S. Department of Agriculture Forest Service, Bugwood.org

Figure 63. Adult female cedar wood wasps (*Syntexis libocedri*, Anaxyelidae) lay their eggs in twigs or stems of recently burned (or even smoldering) incense cedar, red cedar, and juniper. The larvae burrow into the stems. Length ¹/₂ inch (12 mm).

Figure 64. To fit inside an egg, parasitoids must be relatively small or the egg must be relatively large. The egg parasitoid shown here, a wingless internal parasitoid of spider eggs (*Baeus*, Platygastridae), is 1/33 inch (0.7 mm) in length. It is a giant compared to the smallest insect known, a parasite of barklice eggs in which males reach 5/1000 inch (0.14 mm) in length. Photo by Norman F. Johnson

Figure 66. An adult female ensign wasp (*Hyptia harpyoides*, Evaniidae). The larvae of ensign wasps are predatory parasitoids of eggs inside cockroach egg cases. Ensign wasps are sometimes seen flying around in food preparation areas or vending machines in public buildings. It would be advisable to not eat from such places. Length 1/4 inch (6 mm).

Figure 65. Parasitoids that attack barrel-shaped eggs, such as those of stink bugs, are generally short and somewhat squat. *Psix tunetanus* (Platygastridae) is typical of such shapes. Length 1/16 inch (1.5 mm). Photo by Normal F. Johnson

Figure 67. Species of *Inostemma* (Platygastridae) are egg-larval or egg-pupal parasitoids of gall midges, including the blueberry gall midge. The unique projection arising from the first abdominal segment houses the extremely long ovipositor—long at least for a wasp—that measures 1/16 inch (1.5 mm) in length. Photo by Norman F. Johnson

Figure 68. In North America, the biology of the family Gasteruptiidae is unknown, yet the wasp is fairly large (up to 1.5 inches, 40 mm) and not uncommon. In Europe, females insert their ovipositors into a solitary bee or wasp nest entrance to lay their egg. These wasps do not penetrate solid wood as do most of the wasps with long ovipositors that attack wood-boring beetle and sawfly larvae. Photo by Bob Carlson

Figure 69. Silken cocoons of larvae of a braconid wasp (*Cotesia*, Braconidae; left) are often thought to be eggs. The wasp inserted eggs into this tobacco hornworm (*Manduca sexta*), but the wasp larvae have burrowed out from inside the host after eating it alive. The adult of *Cotesia congregata* appears innocent enough, but her offspring tell a different story. Length 1/8 inch (3 mm). Larva photo by Robert L. Anderson, U.S. Department of Agriculture Forest Service, Bugwood.org

Figure 70. Figitid wasps (Figitidae) are larval-pupal parasitoids of flies. They lay their egg in the larva, but adults emerge from the pupal stage. Length 1/4 inch (6 mm). Photo by Matthew Buffington

Figure 71. Ichneumonids are extremely common, though we don't always know what they do. Based on reports of related genera, this *Cryptanura* probably attacks lepidopterous pupae. Although most ichneumonids can't sting (in the sense of the stinging parasitoids), this group can poke really hard, inject venom, and inflict pain that may last several days. Wasps with short, thick ovipositors should be treated with respect. Photo by Bob Carlson

Figure 72. Species of *Aphidius* (Braconidae; left) are specialist parasitoids of aphids, which they may attack from the embryonic to adult stages. Length 1/8 inch (3 mm). Each of the light brown aphids is simply an aphid bag with an *Aphidius* inside.

Figure 73. Species within a genus often have quite different biological requirements. Some members of the genus *Megastigmus* (Torymidae) are parasitoids of gall-forming insects, while others are seed feeders. This wasp, *Megastigmus albifrons*, is an internal feeder in the seeds of nearly a dozen pine species. Length ³/₈ inch (9 mm). Photo by Michael Gates

Figure 75. The California oak apple gall, though often seen, is caused by the wasp *Andricus californicus* (Cynipidae), likely never seen by anyone but entomologists. Length ¹/₄ inch (6 mm).

Figure 74. These globular galls (right) caused by a cynipid wasp (*Atrusca aggregata*, Cynipidae) are among the most intricate of galls even though they appear as simple spheres from the outside. In the center of each gall, a tiny cell hangs suspended by delicate filaments (below). The wasp larva is developing inside.

Figure 77. One of the most colorful of wasps, yet as small as a match-head, the wasp *Tanaostigma stanleyi* (Tanaostigmatidae) causes galls on *Acacia* in the southwestern United States. Length ⅛ inch (3 mm).

Figure 76. Cynipid gall wasps (Cynipidae) form species-specific galls of great variety and location. These acorn-plum galls caused by *Amphibolips quercusjuglans* are a bit unusual even by gall wasp standards. Photo by Steven Katovich, U.S. Department of Agriculture Forest Service, Bugwood.org

Figure 78. The orchidfly (*Eurytoma orchidearum*, Eurytomidae) is actually a wasp. Its larvae live in orchid roots and cause gall-like swellings. Length ⅛ inch (3 mm).

 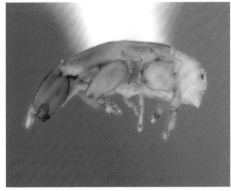

Figure 79. A female fig pollinating wasp (*Pegoscapus assuetus*, Agaonidae; left) appears wasp-like (length $1/32$ inch, 0.8 mm), but her male counterpart looks more like a shrimp (length $1/64$ inch, 0.4 mm). He never leaves the fig, his sole function being to mate with females before they escape.

Figure 80. Wingless female velvet ants (Mutillidae) appear much like real ants (Formicidae) as they crawl over the ground searching for solitary wasp and bee nests to invade. They parasitize the maturing occupant—larva or pupa—when they find one. Males are winged, appearing little like the females of their species. Photo by Carll Goodpasture

Figure 81. The thistledown velvet ant *(Dasy-mutilla gloriosa)*, despite its common name, is less velvety and more punk-like than most other species. This velvet ant is a parasitoid in predatory sand wasp (Crabronidae) nests, where its larvae consume both wasp larvae and provisions. Length ¹/₂ inch (12 mm).

Figure 82. Cuckoo wasps (*Chrysis*, Chrysididae) are metallic green or blue jewels that parasitize bees and wasps by sneaking into the nest as it is being provisioned. This species is *Chrysis nitidula*. Some have the ability to curl into balls for self protection. European species are even more colorful, adding metallic red shades. Length ³/₈ inch (9 mm).

Figure 83. Tiphiid wasps (*Myzinum*, Tiphiidae) burrow into the soil to find scarab beetle larvae, which they parasitize with no attempt at creating a nesting chamber. Photo by Bob Carlson

Figure 84. Scoliid wasps (*Scolia*, Scoliidae) are related to tiphiids, and they also burrow into the soil to find scarab larvae. Unlike her lazy relatives, the scoliid female builds a cell that surrounds the larva once she has parasitzed the host. Photo by Bob Carlson

Figure 85. Sand wasps, *Eremnophila aureonotata* (Sphecidae), here doing what wasps do when the female is not digging and provisioning a nest. Males, in general, are relatively useless otherwise. Females are caterpillar hunters. Despite their slightly fearsome appearance and the fact that they are members of the stinging wasp group, they cannot sting. Photo by Bob Carlson

Figure 86. Females of the predatory wasp genus *Podalonia* (Sphecidae) commonly hunt soil-inhabiting cutworm larvae, which they dig up and transport to a spot where they excavate a nest. This differs from the typical practice of digging the nest prior to hunting. A single larva is placed in a cell, and the wasp lays an egg on its side. The wasp then abandons its young to survive on its own.

Figure 87. A typical sand wasp, *Ammophila* (Sphecidae), with the characteristic elongate petiole of the abdomen and black and red coloration. Species of *Ammophila* catch and paralyze larvae of moths and sawflies, which are placed in cells excavated in the ground. Depending on the wasp species, there are generally one or two cells per nest, and each cell may contain from one to a dozen host larvae.

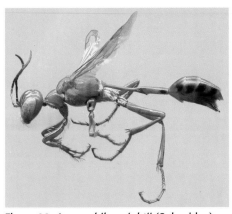

Figure 88. *Ammophila wrightii* (Sphecidae) appears substantially different from the typical red and black sand wasp species. This wasp appears much like a harvester ant when running over the ground and can be found in the same areas. Length ³/₄ inch (18 mm).

Figure 90. Potter wasps, in this case *Eumenes fraternus* (Vespidae), create small mud pots that are attached to almost any object. Each pot receives a caterpillar, and an egg is suspended from the ceiling by a thread. Photo by Bob Carlson

Figure 89. *Monobia quadridens* is a eumenid wasp (Vespidae) that uses abandoned nests of other bees in which to establish its own nest. It provisions with caterpillars. Photo by Bob Carlson

Figure 91. Species of *Cerceris* (Crabronidae) provision their larval cells with adult beetles. This wasp has a weevil tucked under its body. As you can see from the next photograph, *Cerceris* wasps appear almost identical to species of *Philanthus*, to which they are closely related. The latter, however, provision with bees, not beetles. Photo by Carll Goodpasture

Figure 92. This female *Philanthus* (Crabronidae), poised at the entrance to her nest, has just returned from the hunt with an adult bee tucked closely under her body (although it is difficult to see). Often it is possible to gain some idea of a wasp's identity based on what it is carrying home. Photo by Carll Goodpasture

Figure 93. *Tachytes* species (Crabronidae) are common grasshopper hunters that nest underground. Photo by Susan Ellis, Bugwood.org

Wasps of the genus *Sphex* (Sphecidae; Figure 94) generally hunt katydids but will also take crickets. One species even specializes in leaf-rolling crickets (Gryllacrididae). *Sphex* species are ground nesters, and because they are large and common they have been given names such as the great black wasp and great golden digger wasp. They generally provision a single nest with a single prey item.

Other ground-nesting grasshopper hunters include several dozen species of the genus *Prionyx* (Sphecidae). These wasps are noted for standing over their paralyzed prey, grabbing it by the antennae, walking forward, and dragging it over the ground to their nest. I once studied a species of *Prionyx* that nested along the sandy edge of a Florida tidal marsh. Instead of digging its own burrow, this species entered abandoned crab burrows, remodeling them by excavating side chambers and storing their grasshopper victims within these cells.

Unlike the preceding wasps, species of *Isodontia* (Sphecidae; Figure 95), also called grass-carrying wasps, do not nest in the ground. Instead they head for preexisting cavities such as hollow plant stems, abandoned bee burrows, crevices of many kinds, or even the hollow channels in an old window frame. At my old Maryland house I was surprised on several occasions to see a black, thread-waisted wasp approaching the bathroom window trailing a long piece of grass beneath its body. The wasps carry these blades in their mandibles and use them to plug up the openings to provisioned cells and the nest entrance when complete. Grass-carrying wasps provision their cells primarily with crickets (Gryllidae) and katydids (Tettigoniidae).

Cockroach Hunters

The single family Ampulicidae specializes in hunting cockroaches for its larvae. There are nearly 200 species worldwide, but most are tropical. North of Mexico there are only four ampulicid species, none of which hunt domestic roaches, preferring only those found in the natural world. These wasps are less than 1/2 inch (12 mm) in length and rarely seen. The hunting behavior is rather primitive in these wasps, much as for some of the spider wasps (Pompilidae). An adult female stings her victim, weakly paralyzing it, grabs its antenna, and then drags the roach as she walks backward to a suitable preformed niche in which to hide the hapless victim. One roach is provided per egg, which is placed between the middle legs of the host. The niche is then plugged with leaves, grass, and other debris. Some species are known to place their prey in hollow plant stems and partition between them using debris. Some species are also known to bite off part of a roach's antenna and drink its body fluids.

Fly Hunters

Several members of the family Crabronidae collect adult flies as provisions for their nests. Species of the genus *Steniola* prefer bee flies (Bombyliidae), but soldier flies (Stratiomyidae), robber flies (Asilidae), flower flies (Syrphidae), and others are used. Males and females have an interesting habit of clustering on vegetation at night and during storms. Nearly 500 wasps have been reported in a single cluster, but smaller numbers are likely more often the case. Females nest in gravely soil, with the deepest nests reaching about 7 inches (18 cm).

A common North American fly hunter is *Stictia carolina* (Crabronidae), also known as the horse guard because it hunts horse flies (Tabanidae) that gather around stables. These flies are blood suckers, so wild animals are fair game as well. The horse guard also hunts screwworm flies (Calliphoridae), stable flies (Muscidae), and others, as well as horse flies. *Stictia carolina* nests in sandy soil, excavating its nest up to almost 2 feet (60 cm) deep and building only one cell. This species is unusual in that it progressively provisions its nest, providing flies as the larva matures. This may take up to a week or more and estimates of total flies needed per larva range from about 15 to 25. The female closes the nest and leaves before the larva finishes dining and pupates.

The genus *Bembix* (Crabronidae) contains nearly 350 species worldwide, all of which apparently are adult fly hunters. They are often referred to as sand wasps because they nest in sandy areas, sometimes in large aggregations. *Bembix* wasps basically perform their duties as does the horse guard. Although they also hunt tabanids, these are not as commonly taken and they hunt a much broader spectrum of flies. Their nests are progressively provisioned and consist of one or two cells.

Crabro (Crabronidae; Figure 96) is a common genus of about 80 fly-hunting species. They are ground nesters, digging up to 17 inches (45 cm) in depth and constructing up to 15 cells. Unlike the previous species, these are mass provisioners, placing all the flies the larva will eat into a sealed cell. Up to 100 gregarious nests may be found at a single nest site.

Not all fly catchers are ground nesters. The genus *Ectemnius* (Crabronidae, Figure 97) contains more than 180 species worldwide, some of which are ground nesters and some of which nest in twigs, stems, decaying logs, and holes in sound wood. Approximately 25 species occur in North America north of Mexico, and another 20 species are endemic to Hawaii, where most of the ground nesters apparently occur. The twig-nesting species do not require much in the way of a home, as I once found *Ectemnius scaber* nesting in dried parsley stems that had not been removed from the parent plant. Unfortunately the prey of this species was adult flower flies (Syrphidae), whose larvae are extremely valuable in controlling soft-bodied insects in the garden.

Ant Hunters

It has always seemed odd to me that a wasp would specialize in hunting ants, but then nearly everything about Hymenoptera surprises me these days. The ant hunters are found in one small group (a tribe) of the family Crabronidae. There are several genera and a few species. The ant hunters are fairly specific in what they capture.

The genus *Tracheliodes* contains two species in western North America that nest in plant stems. They both prey upon worker ants in the genus *Liometopum*, of which there are only three known species. These ants form large colonies, are aggressive, and have defensive chemical compounds that reportedly smell like blue cheese (Fisher and Cover 2007), yet despite this they are still the only targets of *Tracheliodes*.

Aphilanthops, a genus of four species, two of which are widespread across North America, collects only winged queen ants of the genus *Formica*. These queens can only be hunted during their reproductive cycle and are captured as soon as they land on the ground after their nuptial flight. The wings are removed, and the ant is carried in the legs of the wasp as she flies back to her nest. *Aphilanthops* nest in the ground, descending to 18 inches (45 cm) below ground level, where two or three ants are placed per cell.

An even more remarkable ant hunter is *Clypeadon* (Figure 98), a genus of seven species, all found in the American West. These are ground nesters, excavating to nearly 10 inches (25 cm) deep. Cells may contain from 10 to 20 ants of a specific species. *Clypeadon* females wait near an ant nest entrance, remaining motionless until an ant hurries by. The female then rushes the ant, stings it, and carries it away. Alternatively some female wasps have been seen to rush the ant nest opening, baiting an ant to attack, then running away until the ant is separated from the herd. The wasp then turns around, stings it, and carries it away. Very cunning these hunters! These wasps are equipped with a special ant clamp at the tip of their abdomen in which they clamp the ant, arching it forward between their hind legs for additional support.

Bug and Planthopper Hunters

A large number of predatory wasps hunt members of the order Hemiptera, or true bugs, and all are in the family Crabronidae. Included among host records are leafhoppers (Cicadellidae), treehoppers (Membracidae), delphacids (Delphacidae) and related planthoppers (that is, various Fulgoroidea; Figure 99), aphids (Aphididae), spittlebugs (Cercopidae), assassin bugs (Reduviidae), leaf-footed bugs (Coreidae), and stink bugs (Pentatomidae). Because these bugs have an incomplete metamorphosis, both immature and adults are hunted.

Members of the genus *Pluto* provision their nests with planthoppers and leafhoppers. They nest in the ground, burrowing to about 2½ inches (65 mm) and placing up to 20 prey items per cell. Related genera use aphids as prey. *Hoplisoides* species use adult and nymphal treehoppers, leafhoppers, and planthoppers. They nest in open sandy areas, usually with the entrance concealed beneath a leaf or rock. A short burrow is dug to about 3½ inches (90 mm) and up to three cells with 10 to 15 prey each are provisioned. Species of the genus *Astata* (Figure 100) primarily hunt adult and nymphal stink bugs. They nest both in sandy or hard-packed soils.

Many additional genera of Crabronidae use Hemiptera as hosts, including *Nysson*, which actually acts as a cleptoparasitoid of her relatives. Female *Nysson* lay in wait as the real hunters enter and leave the nest in pursuit of prey. When the opportunity arises, the *Nysson* female enters a nest, finds a cell being provisioned, and lays an egg on the bottommost prey. The nest maker lays an egg on the last provisioned prey (the topmost), and then seals the cell. The *Nysson* egg hatches first, the larva finds and eats the egg of the nest maker, and then dines on the stored prey.

Cicada Killers

The cicada killers, though only a few species, are given their own section because they can become extremely common in any given location, they are big, and people fear them. These wasps belong to the genus *Sphecius* (Crabronidae). There are only four species north of Mexico and a couple dozen throughout the world. The common species east of the Rocky Mountains is *Sphecius speciosus* (Figure 101). West of the Rockies we find *S. grandis* and *S. convallis*, and in Florida, *S. hogardii*.

Cicada killers, being large wasps nearly 2 inches (50 mm) in length, are fairly scary in appearance, but they have a habit that makes them even scarier. Males are territorial and investigate anything that wanders into their territory. They appear to be aggressive, though in fact there is little they can do except bump an opponent. Females can pack quite a sting, but fortunately they are not aggressive. Another aspect that makes cicada killers appear fearsome is they often build up large nesting aggregations in a given area. One huge wasp is fearsome enough, but hundreds can be alarming.

Cicada killers nest in the soil. Females of *Sphecius speciosus* are known to dig long, underground burrows reaching up to 4 feet (1.2 m) with a dozen or so branching runs, each ending in a cell provisioned with one to several cicada adults, depending upon their size. On occasion several females may use the same burrow, each provisioning her own cells but not working together. Cicada killers, then, may be thought of as communal nesters in some situations. Unfortunately for the gardener, the tailings from these excavated burrows end up as a pile of soil at the entrance to the nest. If plants are present, they can become covered and die. In Maryland I had a rock garden in which several dozen cicada killers were nesting, and the soil from these nests smothered small, ground-hugging plants if I did not remove it.

Because cicadas are rather large prey items, they pose a problem even for a large wasp. After stinging and paralyzing the cicada, the wasp holds it lengthwise and upside down within its middle legs, then drags the prey home to its previously excavated burrow. In some cases she may attempt to fly with it, but taking off from ground level can be difficult when so heavily weighted. I've seen cicada killers drag their prey up a tree trunk, then attempt to fly off from a height only to land promptly back on the ground. It seemed quite frustrating to an onlooker, but then whose life isn't from time to time?

Thrips Hunters

At the opposite extreme of cicada killers are those sorts of wasps that prey on extremely tiny hosts. Thrips (Thysanoptera) illustrate the fact that almost any insect can serve as host to the right wasp and that hunting wasps need not be huge. Some hunting wasps are as tiny as parasitoids, being less than $1/8$ inch (3 mm) in length, and so are not likely to be seen attacking cicadas or spiders. In fact, they are not likely to be seen at all. Several such wasps occur within the family Crabronidae, especially the subfamily Pemphredoninae, the same family to which the cicada killers belong. Being small, relatively little is known about thrips-hunting wasps, but in the few known cases females provision with thrips or aphids. Because the life histories of these tiny wasps are sketchy at best, I devote a few sentences to one of the thrips hunters by way of introduction.

Pulverro monticola (Crabronidae; Figure 102, left) is a ground-nesting wasp that provisions its nests with adult thrips. These thrips are captured at flowers and carried back to the wasp's nest in her huge jaws (huge, that is, because they are half as long as the head; Figure 102, right). Little is actually known about this wasp's nest architecture because attempting to excavate a nest a mere $1/16$ inch (1.5 mm) in diameter is no easy task. (I can say this from personal experience, having coauthored the only biological paper on the genus with Richard Bohart, my major professor.) The female digs a burrow in the soil, reaching at least 3 to 4 inches (75 to 100 mm) in depth. Off the main burrow she creates a small cell into which she places about 30 thrips and an egg. Upon completion the cell is blocked with soil, and she moves onward to construct and provision another cell. We could not determine the number of cells for each nest, but the incomplete nest we successfully excavated while still being provisioned by the wasp had more than 60 thrips. To place each thrips takes one hunting sortie, so the female obviously spent much of her time flying back and forth from where she found thrips to where she interred them. In the vicinity, we noted male and female wasps as well as thrips on flowers of *Potentilla*, *Nama*, *Phacelia*, *Ligusticum*, and *Penstemon*. One of the truly amazing things about these tiny wasps is that they not only make dozens upon dozens of flights to and from the nest, but they have to find large numbers of prey even tinier than themselves.

Spider Hunters (Including Mud Daubers)

It is ironic that spiders, which are notorious insect hunters, must contend with an entire family of predatory wasps that turn upon them as prey for their own offspring. These are the spider wasps (Pompilidae). Members of the subfamily Ceropalinae act more as parasitoids in behavior than as predators. They number only about 200 species, as compared to about 4000 species worldwide of the other subfamilies, which are mostly true hunters. Pompilids are often seen walking jerkily over the ground and sometimes may even be seen dragging their spider prey backward (Figure 103). This is considered a primitive behavior because the wasp cannot see where it needs to go. From time to time the wasp may drop its host, survey the terrain, then begin the dragging process again.

Among predatory wasps, pompilids have the simplest forms of behavior of all. They need find only a single spider, lay an egg on it, and then depart. Even their most complex behavior is simple, although it takes a great deal of bravery (or foolishness) to use it. In the simplest form of hunting, spider wasps attack the spider in its own lair, sting it, lay an egg on it, and then depart. Thus, the spider is in its own protected place and the spider wasp does not have to excavate a burrow. It's as if the wasp had consulted a book of hunting for dummies. Of course, the real problem here is that the wasp has to be fearless enough to walk into a spider's den or invite it to attack. One example of this behavior is a genus of spider wasp (*Aporus*) that attacks trap-door spiders. The wasp burrows into the soil near a trap-door spider nest, provoking the spider to exit its nest, at which point the wasp overtakes it, paralyzes it, then drags the spider back to its own nest. The spider's castle then becomes its coffin as the wasp lays an egg, exits, then seals the trap door shut. At the opposite end of the nesting spectrum, the most advanced members of the family (*Auplopus*) build clusters of mud cells, each of which is provisioned with a single spider, as is the case for all known pompilids.

The seriously gigantic tarantula hawks (*Pepsis*; Figure 104), of which about a dozen species live in the American Southwest, are both the most awesome and fearsome of creatures. These are huge wasps, up to 2 inches (50 mm), with iridescent black bodies and red wings. In flight they appear twice as large (or if really close, some multiple of twice) as they really are. Because they attack tarantulas, it might be suspected that they possess a fearsomely painful sting. (I have been stung by their cousins, *Hemipepsis*, which are about half their size. I don't especially recommend it, and I stay the hell away from the real thing.) Tarantula hawks have no predators to speak of, and for good reason. After paralyzing a tarantula, preferably a female because of its greater body mass, *Pepsis* either stuffs the spider back in its burrow or drags it to a prepared nest.

Although Pompilidae is a relatively poorly known group containing thousands of species, there is a single species of spider hunter that is perhaps the most commonly seen of all solitary predatory wasps. In spite of that, people know little about this wasp, the cosmopolitan mud dauber, *Sceliphron cementarium* (Sphecidae; Figure 105). This species also happens to be one of the few wasps that has an entire book written about it, *The Ways of a Mud Dauber* (Shafer 1949). Mud daubers make the lumpy mud nests often seen on the sides of houses, under eaves, and under bridges.

Mud daubers begin their nests by gathering up a small ball of mud in their mandibles. This is carried back to a suitable substrate, where it is applied, layer upon layer, in small, overlapping half arches. The mud cell is begun at the bottom and worked toward the top. Generally a single cell (or tube) can be constructed and stocked with prey in a day, although in some cases one to two weeks have been reported to complete a single cell. Up to two dozen spiders are hunted and stored per cell, an egg is laid (there is debate about whether upon the first or last spider), and the tube is sealed with mud. Once a tube is completed, another is begun next to it, until several such tubes are built and provisioned. The wasp may then coat the collective cells with mud, or not, depending, I suppose, upon the work ethics of each female wasp. The larvae that develop within the cells overwinter until the next year, when they emerge to repeat the cycle (Figure 106). For those that fear such things, a sealed mud dauber nest is in no way harmful to humans. No wasp will rush out to ward off intruders, as in yellow jacket nests.

The genera *Trypoxylon* and *Trypargilum* (Crabronidae) are also exclusively spider hunters. Several species create mud cells, but most use preexisting cavities in which to store their prey. The latter species are sometimes called keyhole wasps because they build their cells in nail holes, keyholes, and even folded newspapers (if left untouched). These wasps do not build the nest of mud, but they do separate the cells with mud partitions. The most well known species within these spider hunters are the pipe organ wasps (Figure 107), which make a series of mud tubes arranged a bit like a panpipe. Each band (or coil) of mud is easily seen in the structure of the nest. As with the mud daubers, spiders are collected, placed in the tubes, and sealed. Unlike the nests of mud daubers, however, a male wasp might stand guard within the entrance to each tube. Males are harmless, of course, but they might be a bit surprising to those unaware of their presence.

Pollen Hunters

Because nothing is either simple or direct within the Hymenoptera, we are faced with the conundrum of predatory wasps that are not predatory but pollen-collecting wasps. That is, they feed their larvae pollen and nectar, just as bees do. These are species of Vespidae found in the subfamily Masarinae (Figure 108), which at one point was considered a family unto itself. It may be logical to ask why pollen wasps are wasps and not bees, and the resolution to this question can be found in the chapter on bees.

Pollen wasps are not common, numbering slightly more than a dozen species in North America and more than 200 worldwide. Depending upon the species, these wasps nest in the ground or make cells of mud or cemented sand particles that they attach to objects such as rocks. Each nest is provisioned with pollen and nectar and sealed.

A Miscellany of Hosts

A very few predatory wasps seem not to care what they collect for their offspring. In most cases, these prey must be alive. For example *Lindenius columbianus* (Crabronidae) will take flies, bugs, and even parasitic wasps to provision its nests. *Glenostictia scitula* (Crabronidae), a species native to the American Southwest, is reported to use 43 species of bees, predatory wasps, parasitic wasps, true bugs, and flies in 27 families of insects.

There are predatory wasps that act as scavengers, collecting dead and nearly dead insects and spiders they find along the way. For example, *Microbembex monodonta* (Crabronidae) gathers hosts in nearly a dozen orders of insects as well as spiders, all of which are dead and none of which can fight back. It must be said that these are more the oddball sorts of wasps, and most solitary hunters are fairly specific in the prey they hunt to stock their larval cells. The social predatory wasps are much less demanding and much more generalists in what they hunt.

Social (Eusocial) Hunters

Social predatory wasps, more properly called eusocial wasps, are commonly known as yellow jackets, hornets, and paper wasps. All are members of the family Vespidae and subfamilies Vespinae (yellow jackets, hornets) and Polistinae (paper wasps). As noted from the preceding discussion, not all vespids are social. In fact, there are roughly five times more solitary species, but living alone and not being the least bit aggressive, these recluses are overshadowed by their colonial cousins.

The basic living unit of a social wasp larva is an inverted papier-mâché comb created from chewed plant fibers of many kinds (Figure 109). It is not uncommon to hear and see paper wasps tearing tiny wood splinters from unpainted wooden fences made of rough cut lumber. Hornets house their combs in structures that appear to be paper bags and are generally found hanging in trees or shrubs (Figure 110), though some prefer enclosed spaces. Most paper wasps hang their combs openly for all to see, but they are rarely seen anywhere other than under the eaves of homes. And yellow jackets generally hide their combs underground or in an enclosed space (Figure 111). They can, however, make aboveground or underground nests that are enclosed, as with hornets.

The general life cycle of any of these wasps is a yearly one, with an overwintering queen beginning a new colony in the spring and an ever-increasing buildup of daughters throughout the summer. Production of reproductive males and new queens begins in autumn. After mating, the new queens overwinter and the old colony dies. This is temperature dependent, so that in warmer winters in cold climes, colonies may survive a winter. In warmer, southerly areas, colonies may extend from year to year, building up to unpleasantly large and dangerous proportions. I've seen pictures of the entire cab of an abandoned pickup filled to overflowing with a yellow jacket nest, as well as multiple colonies approaching 12 feet (3.6 m) in height wrapped around tree trunks.

As hunters, all these wasps excel. They are powerful, vicious, and not too demanding about what they hunt. Although soft-bodied prey such as caterpillars and sawfly larvae are among the most commonly taken hosts, almost any moving insect can be attacked. I once saw piles of swallowtail butterfly wings lying beneath a butterfly bush and simply assumed that some praying mantids were at work. Then I saw a bald-faced hornet tear the wings off a swallowtail and begin feasting on its body. There were enough fresh wings to tell me that the hornet was using my butterfly bush as her hunting grounds. The wasp masticated the body in preparation for her return home, whereupon she would regurgitate the pabulum to the communal young residing in the cells of her nest.

Social wasps are progressive provisioners that feed their young as they develop. The queen does all the egg laying, at least in theory, whereas the workers do the nest building, cleaning, hunting, and feeding. Vespid wasp queens rule by domination, not by chemical pheromones as in ants and honey bees. Worker females are bullied by the queen, which results in a worker's reproductive apparatus remaining underdeveloped. If the queen should die, these workers can sometimes become queens and lay eggs.

Within vespids, the most problematic to humans are the yellow jackets (*Vespula*; Figure 112). They can be unusually common, uncommonly aggressive, and possessed of gang-like intensity, especially toward autumn when colonies have reached their largest populations and food is becoming scarce. It's at this point that workers, desperate for food, begin investigating beer cans, soda bottles, hot dogs, hamburgers, and chicken legs. They also investigate these things as you put them in your mouth, so it is wise to be very careful when eating. I once pulled into a picnic ground that was eerily empty of picnickers. Upon emerging from the car I discovered why. The yellow jackets were so numerous as to make it impossible even to walk to a table, much less sit at one. It's important to know when you're not wanted and act accordingly.

Next to yellow jackets, hornets (*Dolichovespula* and *Vespa*) seem downright tame. Individuals are quite scary, to be certain, but they don't seem to hang out in gangs like the yellow jackets do. Single hornets rarely charge into your mouth after a bit of chicken as do yellow jackets. Of course, disturbing the nest is absolutely forbidden, but only fools and children would do such a thing, so the rest of us are safe. Although the black-and-white bald-faced hornet (*Dolichovespula maculata*) is the most commonly seen in North America, the European hornet (*Vespa crabro*; Figure 113) is becoming more common, at least in the northeastern United States, and may soon be coming to a playground near you. The yellow, red, and black European hornets are huge, especially when buzzing near you, but I have not noticed them to be overly aggressive. Bald-faced hornets build their combs in paper structures that are usually hung in shrubs or trees, whereas European hornets nest in cavities and concealed places (including attics and sheds), although they still make a paper nest. Another difference is that European hornets fly both day and night and are attracted to lights. If there are large hornets buzzing around your porch lights, these are the ones. In Europe these wasps often attack honey bees but do not depend upon them for food. The Asian giant hornet (*Vespa mandarina*) actually kills off entire honey bee colonies in their search for food.

The widespread paper wasps (*Polistes*, Vespidae; Figure 114) are the most common social wasps found attaching themselves to our homes. There are nearly two dozen species in the United States that make the upside-down, open-faced combs frequently seen suspended from beneath the eaves. These combs rarely get very large, and so the nest mates are not terribly numerous. Among social wasps, these are the most docile of them all, allowing a person to stick their face right up to the nest before showing defensive warning signs, which are wings back (V-shaped), front legs apart, and an intense stare-down. The next stage is not a direct frontal attack, as would happen with yellow jackets or hornets, but a circular reconnaissance flight to assess the situation.

Paper wasps have a couple of commonly occurring cousins in the American Southwest that superficially look like them but actually are fairly distinctive when seen close up. Because these wasps, which belong to the genus *Mischocyttarus* (Figure 115), build the same sort of nests as *Polistes* they are also called paper wasps. *Mischocyttarus* wasps have a long, thin connection between thorax and abdomen that distinguishes them from *Polistes*. They seem to differ in preferred housing areas, and although they nest under eaves, they also nest within structures (exterior fan housings, for example) or in rooms that might remain open during the day.

Some 30 or 40 years ago a species was introduced from Europe into North America that looks more like a yellow jacket than a paper wasp. The European paper wasp (*Polistes dominula*; Figure 116) has made itself right at home, having spread from coast to coast in the northern half of the United States and into Canada, and it was recently found in Hawaii. *Polistes dominula* is termed an invasive species because it is replacing native species of *Polistes* and is interfering with cavity-nesting birds. This species prefers to nest in protected places, including as I once discovered, mailboxes. It also likes bird houses. Unfortunately, as well as looking like a yellow jacket it does not have the docile demeanor of other paper wasps, but rather acts like a yellow jacket. Also, unlike its fellow caterpillar hunters, this wasp is more catholic in its hunting tastes and lays waste to anything it finds. Therefore, unlike other Hymenoptera, you have my permission to dispose of the European paper wasp whenever you find it.

As with the eusocial bees, there are some eusocial wasps that do not play quite fair when it comes to members of their own kind. That is, they are social parasitoids, choosing to infiltrate the established nest of a closely related species rather than build their own. So closely do social parasitoids resemble their victims that their existence was not suspected until the middle of the last century. The infiltrators are found in about a dozen species of paper wasps (*Polistes*), yellow jackets (*Vespula*), and hornets (*Dolichovespula*). Obligate parasitoid species can exist only if they invade the nest of another species, whereas facultative species can either build their own nests or invade those of other species. In the typical situation, a parasitoid female will enter a nest and slowly become the dominant queen by aggressively overwhelming the occupants, including the queen. She may kill the queen outright or allow her to continue reproducing for a while. Parasitoids produce only queen and male progeny, so eventually the original occupants are replaced by the squatters, each of which, once mated, overwinter as an adult and repeats the entire process the next year.

8

The Garden's Pollinators: Bees

MENTION THE WORD *bee* and our perspective on the subject immediately turns to the honey bee. On a good day we might even follow this with thoughts of bumble bees or carpenter bees. At that point most of us will have run out of any sorts of thoughts about bees at all. But what of the resin bees, don't they deserve our thoughts as well? Or the mason bees and sweat bees? The leafcutter bees? The cuckoo bees? The mining and digger bees? The green bees and yellow-faced bees? The squash and gourd bees? The plasterer, polyester, or cellophane bees? The carder bees? And my favorite moniker of all, the shaggy fuzzyfoot bee? There are bees galore and many more, and some deserve a bit of attention now and then, so in this chapter we'll examine aspects of bee life that may prove interesting, if not useful, to the gardener.

Bees, it is currently hypothesized, are simply pollen-collecting wasps. That is, some predatory wasps began to collect pollen instead of insect prey to feed their young. The pollen wasps (subfamily Masarinae) are an example of a possible intermediary step in the road to beedom. The main difference between bees and wasps (as well as other Hymenoptera) is that the bees have branched or feathery hairs and the wasps have simple hairs. During this evolutionary transition, it is believed that the unmodified body hairs of wasps became branched to better trap pollen grains. In addition, structures associated with the legs became modified to scrape and consolidate pollen from these hairs. In some cases the pollen was formed into packets and stored in the corbicula (a sort of pollen basket) on the hind leg, and in others, pollen accumulated in hairs on the underside of the abdomen. Transporting one grain of pollen at a time would have been rather time consuming, so consolidating and transporting as much pollen as possible soon became the imperative. Thus, when you get right down to basics, bees are just fuzzy wasps.

Compared to predatory wasps, bees are much less diverse in what they feed their young: in every case pollen and nectar and rarely plant oils. If we ignore the honey bee, which has about as complex a life as is possible, the basic life of a bee is arguably less complex than that of a predatory wasp. Within bees there are levels of sociality from solitary to eusocial, but analogous behaviors may be found in the wasps. Some bee species are thieves (that is, cleptoparasitoids) of other bees, in which case their larvae eat pollen provisions given to the host bee larvae, thus inadvertently starving the host larvae. This behavior is also found in predatory wasps, though the food is other insects, not pollen. And there are a few colonial bee species that are social (or brood) parasites, invading a host colony and living among the worker bees as if they belonged there—another behavior found in social predatory wasps.

As with predatory wasps, the majority of bees are solitary, building and provisioning nests by themselves. Solitary bees are much less well known than the eusocial kinds (that is, honey bees and bumble bees). Of all the solitary sorts, the large carpenter bee (*Xylocopa*) is likely to be the most readily recognized member of the group. The leafcutter bees are more likely noted by what they do than their actual appearance. They are responsible for the circular cutouts in leaves of roses and other plants. Other than these few examples, there is a tremendous variety of lesser-known characters about which much can be said.

Social bees are defined by the presence of two or more individuals occupying the same nest. Unfortunately this causes some confusion when attempting to segregate bees into groups of behaviors we might find convenient. Therefore, most of us likely think of social (that is, eusocial) bees only in terms of honey bees or bumble bees in which many individuals work together and are directed by the queen. In temperate areas of the world such as the United States and Europe, members of the true eusocial habit are limited to the two groups just mentioned, but in tropical regions there are other groups of similarly social bees called stingless bees (entirely eusocial) and orchid bees (some species eusocial). There are various levels of sociality within bee groupings, including communal, quasisocial, semisocial, subsocial, and eusocial (that is, honey bees, bumble bees), each a little more complicated than the preceding. The distinctions, however, between these categories are not very clear, even to specialists, and species within families and even within genera may be found in several of these categories. Rather than get bogged down in hairsplitting minutia, I'll mention only a few of the more straightforward examples of social behavior likely to be found in the bees in or around our gardens.

To put our woeful ignorance of bees somewhat in perspective, as of October 2008 several checklists of all the valid bee species names in the world placed the number between 19,342 (Discover Life, Apoidea species) and 19,511 (World Bee Checklist; see the table of hymenopteran families). As a result of these overwhelming numbers and our obsessive preoccupation with the honey bee, the behavior of most bees, including even the common bumble bee, remains largely unknown to the general public. In discussing the various groups of bees below, I include numbers simply to give the gardener some idea of what is taking place in nature when we are not looking. These figures are taken from the Discover Life site because it is interactive and the easier source from which to extract the necessary information.

Six families of bees are fairly widespread throughout the world, with a seventh (Stenotritidae) found only in Australia (Michener 2007). Of the world's bee species, roughly one-third (that is, some 6800) are found in North America (Michener 2007). There is no easy way to organize bees into coherent groups based on their biology, shapes, sizes, or degree of sociality, so I've opted simply to discuss them by family unit since there are so few.

I'll also mention some useful additional sources of information about bees as we progress through the families. As you can imagine, hundreds of books, both technical and popular, have been written about the honey bee, and a search at any internet store selling new or used books will turn up more titles than is humanly possible to read. Conversely there are many fewer books on bees in general, and these tend to be technical, terribly expensive, and tending toward the economics or taxonomy of specific groups. By far, the most comprehensive book on bees, at 953 pages, is Charles Michener's *The Bees of the World* (2007), which summarizes everything we know about bees and includes about 2500 references for those who wish to delve further into the subject. I have gleaned much useful information from this source. Unfortunately, it is expensive, highly technical, and largely taxonomic, though it does encapsulate the biology of every group known to humankind. A much more user-friendly overview of bees from a biological standpoint is *Bees of the World* by O'Toole and Raw (1999). For diehard bee fans there is even an annual workshop called "The Bee Course" held at the American Museum of Natural History's Southwestern Research Station in southeastern Arizona (see the list of useful websites).

Family Megachilidae: Leafcutter Bees, Mason Bees, Resin Bees, Carder Bees

Although the family is commonly referred to as leafcutter (or leafcutting) bees, this name describes but a small part of the varied habits of this family, of which there are nearly 4000 species in the world and more than 850 in North America. Megachilid bees are unique because they carry pollen in bands of hairs on the ventral surface of their abdomen, not on their hind legs as do other bees (unless they are cleptoparasitoids, which don't carry pollen at all). Megachilids are among the most useful pollinators found in gardens, although some species come with a negative trait as suggested by the family's common name. Unlike other bees, which secrete linings for the larval cells they construct, megachilids carry foreign materials to do the job. They should perhaps be referred to commonly as builder bees. The materials, which are carried within jaws or legs, are used to line cells, create partitions between cells, or construct entire cells.

The biology of megachilid bees is somewhat simple, depending upon how one looks at it. Given the family as a whole, it is simple in that most species are usurpers. That is, they do not excavate their own nests, but occupy cavities of every imaginable kind in soil, rock crevices, holes in buildings, and abandoned burrows in dead twigs, tree limbs, or pithy stems created by other insects. Within these spaces, megachilids construct one or more cells—analogous to self-contained barrels—using various materials as discussed below. Each cell is provided with pollen, nectar, and an egg. The cells are generally arranged in a linear fashion, one cell next to another in a row, but some species place their cells one atop the other or side by side. Generally there is one generation per year, with the larvae overwintering as prepupae and emerging the next spring. Nothing too complicated there. The complicated part arises because there are many species, and cell construction is accomplished using different sorts of materials to line or build them. For this reason, megachilid bees have gained a number of colloquial names, based largely on facets associated with a chosen cell lining or nest construction. The following examples should make my point.

The name leafcutter bee is generally given to the genus *Megachile*, of which there are more than 200 species in North America and nearly 1200 in the world. (Some species of *Megachile*, however, are called resin bees, which are discussed below.) These are the bees that cut circular patterns in leaf margins or flower petals in some cases (Figure 117). The surgically removed piece of leaf is taken back to the nest and used to line the cells being constructed for each bee larva. It takes many such circles to create a single cell. With few exceptions, leafcutters do not excavate their own nest, but find hollow twigs, abandoned beetle borings in trees, or the linear spaces found between clapboards and sidings on houses, barns, or sheds. They readily accept holes drilled in wooden blocks, hollow bamboo poles, or even bundles of soda straws. A few uncommon *Megachile* species nest in soil, but they are the renegades of the genus. Within the chosen tubular structure, a series of cells is laid end to end, each lined with leaves plastered together to form a small chamber. Each chamber receives a pollen-and-nectar mass and an egg, then is capped off using bits of leaf. The cell cap then forms the bottom of the next cell, which is constructed the same as the first. Up to a dozen cells are formed in this manner. Of the many species of leafcutters, the name most likely known to the public is the alfalfa leafcutter bee (*Megachile rotundata*), which is a Eurasian native introduced into the United States especially to provide pollination for the production of alfalfa seed.

Next to leafcutter bees, mason bees (*Osmia*) are likely the most common megachilids to be found in the garden. Some 150 species are known from North America and just over 300 in the world. These bees nest in many different places, including soil, wood, hollow objects, and cells merely attached to objects. Mason bees do not line their cells with leaves but simply place their pollen-and-nectar mass in the tubular nesting cavity, then partition it off with a wall of soil particles mixed with saliva—thus their masonry moniker. Often these bees are noticed more by their work than by their presence. If the gardener sees a hole in the side of a shed, for example, that has been plugged with mud, it is most likely the result of a mason bee. On my porch sits a bench made up of various-sized bamboo poles. In every open-ended pole of suitable diameter, there is now a mason bee nest. I only discovered this because I briefly saw a bee hovering near the end of one of the bench's poles. Other hollow objects are sometimes used by *Osmia*, including a European species (*Osmia bicolor*) known to nest in empty snail shells.

Several mason bees are recognized for their excellent pollinating work, including the hornfaced bee (*Osmia cornifrons*), blueberry bee (*O. ribifloris*; Figure 118), Maine blueberry bee (*O. atriventris*), blackberry or raspberry bee (*O. aglaia*), and orchard mason (or blue orchard) bee (*O. lignaria*). The hornfaced bee was introduced from Japan into North America to pollinate apple trees, which an individual bee can do at a rate of 2000 flowers per day. One of the blueberry bees (*O. ribifloris*) is a native of the western United States that normally pollinates manzanita, but can be used on blueberries, whereas the Maine blueberry bee is an eastern U.S. species that is being tested for blueberry pollination. The raspberry bee is native to the Pacific Northwest. The orchard mason bee, another native, has quite a following among those seeking pollinator services and even has an entire book devoted to it (Bosch and Kemp 2001). Apparently, two or three orchard mason bee females can pollinate an entire apple tree, and they can double cherry production compared to honey bees. They are especially valuable because they work better than honey bees in bad weather.

Resin bees are another group of megachilid bees, with about 40 species of *Anthidiellum* and *Dianthidium* in North America (about 70 world species) and a few species of *Megachile*. These bees harvest tiny bits of plant and tree resin with which to build their cells. Sand, quartz grains, or tiny pebbles may be incorporated into the resin. In some cases the end result is a hardened, cement-like pot that is attached to objects such as rocks, twigs, or leaves. Single or multiple cells can be constructed, depending on the species. Resin bees are not particularly common, but in the early 1990s the so-called giant resin bee (*Megachile sculpturalis*) was introduced along the east coast of the United States from Asia and has now spread throughout the southeastern states. This nearly hairless, blackish bee is about 3/4 inch (20 mm) in length and uses preexisting cavities in wood, such as old carpenter bee burrows. Although females can sting, these bees are antisocial, preferring flight to fight. The term *giant* used for this bee is truly a misnomer, however, as the Indonesian species *Megachile pluto* is twice as large, at 1 1/2 inches (40 mm), and is among the largest bees in the world. It builds its resin nests in termite mounds.

Finally, there are the carder (or cotton) bees (*Anthidium*), with some 30 species in North America and about 100 worldwide. Carder bees collect plant fibers and mix them with saliva to form a nest-like, combed-cottony cell in which they place nectar, pollen, and an egg. These cells are placed end to end in a natural crevice or cavity or an artificial one such as an opening in a window frame. The wool carder bee (*Anthidium manicatum*; Figure 119) was introduced from Europe into North America for its pollinating efficiency, especially alfalfa grown for seed. Carder bees are extremely agile, fast flyers, often hovering around a flower before alighting on it. They appear more like a wasp than a bee, with combinations of striking black and yellow coloring.

Although most megachilids are solitary bees, a few have adopted the communal habit. Some genera are cleptoparasitoids of other genera within their own family. The common strategy is to enter a working bee's nest and lay an egg in a cell being constructed or provisioned by the host bee. In some of these cleptoparasitoids the adult female will kill the host egg or larva, whereas in others the larva of the parasitoid does the deed. Some of these bees, such as the genus *Coelioxys* (Figure 120), appear nearly identical to the *Megachile* bees they attack, and it is difficult to tell the pollinator from the cleptoparasitoid. In this case the cleptoparasitoid has no pollen-storing areas beneath its abdomen, as does the pollinator, because females have no need to collect pollen. In at least one genus (*Stelis*), some species chew their way into the hardened cell of a resin bee, lay their egg, and then seal the cell with identical resin.

Family Andrenidae: Mining Bees

There are nearly 3000 species of mining bees in the world, of which nearly 1500 occur in North America. Of these, two-thirds belong to two common genera, *Andrena* with about 1400 species and *Perdita* with more than 600 species. The latter genus occurs only in North America. Species of *Andrena* range in size from about ¼ to ¾ inch (6 to 18 mm) and are generally black or dark, with white or gray hair bands on the abdomen. *Perdita* (Figure 121) are among the tiniest of all bees, ranging in size from ¹⁄₁₆ to ⅜ inch (1.5 to 9 mm). They are nearly hairless, pallid in color, and generally have yellow bands on the abdomen. Many species of mining bees restrict their pollen collecting to specific plants, such as a single genus (for example, *Viola*), or more commonly related plants in families such as Polemoniaceae, Fabaceae, Onagraceae, Asteraceae, or Euphorbiaceae.

As might be surmised from their common name, mining bees nest in soil, using either flat or banked surfaces. They dig their own nest, which is generally vertical with lateral branches each ending in one or two cells. Andrenids generally line their larval cells with a waxy substance produced in the abdomen, but some *Perdita* do not. Species of *Perdita* prefer to nest in sand and are most commonly found in the southwestern United States and Mexico. Andrenid bees generally have one generation per year, with the larvae overwintering as prepupae or pupae that emerge as adults the next spring. As with some species in other families (for example, Colletidae, Halictidae), large nest aggregations may appear after several years when females nest in a favorable habitat. In some species, female andrenids are known to use the same nest entrance, often in large numbers, and so are considered communal bees. No species are yet reported to be cleptoparasitoids. At least one species, *Perdita floridensis*, can survive in nesting sites that are submerged under water for several months during the winter.

Family Colletidae: Plasterer Bees, Yellow-Faced Bees

More than 300 species of colletid bees occur in North America, with nearly 2500 throughout the world. Plasterer, polyester, or cellophane bees (*Colletes*), representing one-third of the family, are typically dark, hairy bees, with white hair bands on their abdomen. They range in length from roughly 1/4 to 3/4 inch (6 to 18 mm) in length. Plasterer bees dig horizontal burrows in earthen banks (rarely using abandoned burrows) and often form large nesting aggregations. These burrows may reach 4 to 5 inches (10 to 12 cm) in length. Colletid females do not carry pollen on their legs, but return to their nest with pollen housed internally in the crop. Plasterer bees begat their name based on the habit of lining brood cells with a watertight substance, which is secreted from the abdomen and is reminiscent of cellophane. Within each bag-shaped cell is placed pollen, nectar, and an egg. About 10 cells are provisioned and the nest closed.

The related yellow-faced bees (*Hylaeus*; Figure 122), representing another third of the family, primarily nest in twigs or stems, again lining each cell with a cellophane-like material. A few species are known to nest in preexisting cavities, including beetle burrows, nail holes, or even abandoned plant galls. These bees are the same size as *Colletes*, black, and nearly hairless and have yellow markings on their face. Because pollen loads are internal, as with *Colletes*, it has been somewhat difficult to determine exactly what these bees are pollinating, but at least three North American species are known to pollinate rosaceous plants. In both genera, there is generally one or two generations per year, and the larvae overwinter to pupate and emerge as adults the next spring or early summer.

Family Melittidae: Oil-Collecting Bees

This is a small family of small, rarely seen bees. I mention them only in the interest of completeness as the status of the family is somewhat debatable. (For those who demand precision in all things, the family name Melittidae applied to species found in North America may actually refer to bees both in the family Melittidae and Dasypodaidae. For those who don't, forget you ever read this.) Fewer than 200 species are known worldwide, with about 30 found in North America. Roughly half of these occur in the United States and Canada; although some species are widespread within this area, they generally are inhabitants of warmer, xeric habitats. North American species vary in size from about $1/8$ to $5/8$ inch (3 to 15 mm), are dark colored, and nest in the soil. Their nests are composed of long burrows ending in one or two cells. Species of the most common genus (*Macropis*) are called oil-collecting bees because they collect oil, not nectar, to mix with pollen to provision their young. They also use this oil to line the brood cells. So far only oil from flowers of loosestrife (*Lysimachia*) is known to be used. Another unusual aspect of the North American genera is that they are most closely related to genera in Africa.

Family Halictidae: Sweat Bees, Little Green Bees

Nearly 4000 species of sweat bees are known from virtually all regions of the world, and of these more than 800 are represented in North America. Among bee families, halictids display the broadest range of biology and behavior next to the Apidae. They are the most behaviorally complex of all the so-called solitary bees, ranging from solitary to communal, semisocial, and primitively eusocial (O'Toole and Raw 1999). In addition, among the huge number of species some are cleptoparasitoids and some are social (brood) parasitoids.

Not all sweat bees are sweat bees, if you take my meaning. Typical sweat bees (*Lasioglossum*) derive their name from a tendency to alight on our bare arms or necks as we garden in the hot sun. Presumably these bees are seeking salts and moisture from our perspiration. They are black with slight metallic tinges and are small, mostly in the $1/4$ to $5/8$ inch (6 to 15 mm) range. Their cousins, however, the little green bees, are about the same size but have less predilection for sweat and are corresponding more beautiful. They are highly metallic green (for example, *Augochlorella* and *Augochlora*) or metallic green with yellow-striped abdomens and legs (*Agapostemon*; Figure 123). I call them "little green bees" because they have no common name and they are little and green, which seems to fit better than the

name sweat bee. In South and Central America there are many halictid little green bees as well as other green bees that qualify as sweat-
aholics, but the latter are orchid bees in the family Apidae.

Although some halictid bees nest in rotten wood, the majority are
soil nesters, preferring flat, open areas or banks. They dig their own
nests, often forming large aggregations of adults in a suitable area.
Adding to the size of the aggregations is the fact that many species
construct large numbers of cells—hundreds in some species—in
each nest. A long, central burrow is excavated, from which numerous
lateral burrows each lead to one or two cells. These cells are lined
with a waxy substance produced from a gland in the abdomen. The
lateral burrows are usually backfilled with soil so that at any one time
only one branch will be worked by the female bee. Most halictid bees
overwinter as solitary adults in protected places, much as do bumble
bee and vespid queens.

Halictids have a wide variety of social interactions depending
upon the species and sometimes even within the same species. In
nests of some species several egg-laying queens may be present and
generations of bees may overlap. Nonlaying nest mates function as
workers, and there may even be a division of labor among them. In
some of these species, production of males is limited to autumn.
Thus, some halictid bee species may be seen as exhibiting rudimen-
tary forms of the complex eusociality found in honey bees and bumble bees. But it is even more complicated than that because a single
species may display a completely solitary lifestyle in one area (for example, montane or northerly regions), whereas in other regions
(lower or southerly regions) they display some form of sociality.

More than 400 species of sweat bees (*Lasioglossum*) occur in
North America, with about 1700 species throughout the world. The
world's premiere bee researcher, Charles Michener, has referred to
Lasioglossum as "an enormous genus of morphologically monotonously similar bees" (Michener 2007). Not much of an endorsement!
Given the fact that *Lasioglossum* is one of the largest of all bee genera, I've never once found them to be a bother either in the garden or
in the field when I'm collecting insects.

There are considerably fewer species of little green bees (*Augochlorella, Augochlora, Augochloropsis, Agapostemon*) than sweat bees,
though they are much more conspicuous when seen, due to their

brilliant metallic coloration. The total number of world species is slightly more than 300, with just over 100 occurring in North America. Basically, they differ little from the previous group other than the fact that they are not monotonously similar.

The halictid bee family also includes cleptoparasitoids that attack other bees. Some aspects of this behavior, as well as rare instances of social parasitism, are found randomly in species throughout the family and include some that attack close relatives. Species in the genus *Sphecodes*, however, are wholly confined to cleptoparasitic behavior, attacking both distantly related members of its own family as well as genera of bees in the mining bee and plasterer bee families (Andrenidae and Colletidae, respectively). This genus numbers more than 300 species worldwide, with about 90 occurring in North America. Adult *Sphecodes* are rather striking in color with a black head and thorax but a red abdomen. They resemble some species of the predatory wasp family Crabronidae (for example, *Tachysphex*). In most species of *Sphecodes* the adult female destroys the host egg and replaces it with her own, whereas in many other cleptoparasitoids the larva kills or outcompetes its host egg or larva.

Family Apidae: Carpenter Bees, Digger Bees, Cuckoo Bees, Bumble Bees, and Honey Bees

The family Apidae is well known for its social members (honey bees, bumble bees), but less so for its solitary members. The group is both unfortunately large in terms of numbers and a bit of a nomenclatural quagmire, but we shall largely ignore that aspect of the discussion. Currently nearly 1700 species of Apidae are reported from North America, with about 5700 known throughout the world. This figure includes members of the former family Anthophoridae, which now is considered part of Apidae and makes discussions about these bees somewhat more difficult than it ought to be. For clarity, I treat several groups of bees as subsections of the family. Almost all categories of bee sociality, from solitary to eusocial, are covered in this single family, not to mention extreme differences in nest architecture, behavior, and biology. Some species also usurp the pollen stores of other bees and thus are not pollen collectors themselves. In the case of solitary bees (such as digger bees or squash bees), the culprits are called cleptoparasitoids or cuckoo bees. In the case of social bees (such as bumble bees), the phenomenon is called social (or brood) parasitism.

Carpenter Bees, Both Large and Small

Most gardeners are probably familiar with large carpenter bees (*Xylocopa*), which are quite distinctive in appearance. Most appear to be black, bald bumble bees, quite shiny in appearance. Much less familiar are the small carpenter bees (*Ceratina*) that appear nothing at all like their giant cousins. Currently carpenter bees are placed as the most primitive members of the family, though their exact position relative to other bees is not yet writ in stone.

Beginning at the less familiar and smaller end of the duo, there are more than 300 small carpenter bees in the world (Figure 124), with 71 North American species (about 20 species in the United States and Canada). They range in size from 1/8 to 1/2 inch (3 to 12 mm), appearing black with a metallic blue or green sheen. Females of some species have a yellow stripe in the middle of the lower half of the face. All nest in pithy stems of woody plants, which must be broken so that the ends are exposed. The female bee chews into the stem's pithy interior, extruding bits of plant tissue (sawdust) as she progresses. The cells are placed in a linear fashion, one atop the other, and are unlined, containing no gathered materials such as leaves (as in leafcutter bees) or secreted substances (as in plasterer bees). The partitions between cells are made from tightly packed, chewed pith.

More than 400 species of large carpenter bees (Figure 125) are known throughout much of the warmer regions of the world, but the biology of most is unknown. An Asian carpenter bee is credited with laying the largest egg of any insect, some 5/8 inch (15 mm) in length and nearly 1/8 inch (3 mm) in diameter (Vicidomini 2005). In North America there are 39 species, only seven of which occur in the United States, and so far none have been reported from Canada. Most occur in Mexico. Large carpenter bees are about 1 inch (25 mm) in length, the females being entirely black or black with metallic blue or green tinges. In some species the male has a yellow face, and in one species from the American Southwest males are entirely covered with straw-yellow hairs, but the body color is actually dark as in the female. In another species from the southern United States, the males have pale reddish orange hairs on the front half of the body.

In our gardens, these bees are often noticed because of the males, which are territorial and appear to be attacking us defenseless humans face to face. In fact they can do no harm and are basically bluffing their way through life when it comes to confronting large, blundering animals. Unlike small carpenter bees that nest in relatively narrow twigs, most large carpenter bees need a substantial piece of wood in which to bore. Many species nest in solid wood, stumps, logs, or dead trees, but a few adopt the habits of their smaller cousins and nest in plant stems. These have to be stems of a sizeable nature, such as old agave stalks. Unfortunately for humans, they do not object to using lumber that might be associated with our houses, car ports, sheds, or barns. Apparently carpenter bees won't burrow into painted wood, but unpainted wood is just fine. They will even burrow into chemically treated lumber, though this is not likely their first choice. Carpenter bees do not eat the wood, but chew it, gradually pushing the sawdust-like material out the nest entrance or using it to form partitions between cells.

In the United States, the common eastern carpenter bee (*Xylocopa virginica*) is perhaps the best studied of the carpenter bees. Females bore into wood, hollowing out a channel about 1/2 inch (12 mm) in diameter and 12 inches (30 cm) or more in length. They burrow with the wood's grain, but to gain access to the grain they may first have to burrow perpendicular to it. For example in a 4- by 4-inch (10- by 10-cm) pressure-treated post, I have seen them bore through the side, then make a right-angled turn to proceed with the grain. I've also seen them bore into 2- by 12-inch (5- by 30-cm) wood from the side. Obviously the female must make a right-angled turn before she bores through or she would pop out the other side of the board. She likely uses some change in sound to know when to initiate the turn. At the end of the burrow the female places a ball of pollen and regurgitated nectar, upon which an egg is laid. A space is left for development of the egg to adulthood, and a partition cap made of chewed wood is placed over the cell. The cap then acts as the base for the next load of pollen and an egg. This is repeated until 6 to 10 end-to-end cells are completed. Females complete but a single nest in the north, but where it's warmer, several separate nests may be built. Carpenter bees overwinter as pupae or adults (accounts vary) within their nests, emerging in the spring to mate. Nests are reused from year to year by different generations, and new nests are built by an ever-increasing population of bees. Eventually an aggregation of many nests may be found. A friend of mine has a perennial aggregation nesting site in the post of a wooden railroad wigwag that he's set up in his yard.

Some carpenter bees in other regions of the world are reported to be somewhat social in that several females will use the same nest, one or more females will act as guard bees at the entrance, and there is some overlap of generations. This behavior is not yet known in North American species.

Digger Bees and Squash Bees

Prior to being placed in Apidae, the digger bees and squash bees were once all placed in the family Anthophoridae, which shows their close relatedness to each other. As might be expected from the common name digger bee, most members of this group nest in the soil, excavating nests by use of their mandibles. These are primarily solitary bees that do not gather external building materials for nesting. Instead their cell walls are coated with secretions from an abdominal gland, which produces somewhat of a waxy or varnish-like final appearance. Females nest in vertical soil surfaces such as earthen banks, cliff faces, or adobe walls as well as in flat areas. Nest aggregations are common, and occasionally females may use the same burrow entrance, but each maintains her own cell (that is, they can be communal nesters). Many species create small mounds of earth or mud turrets resembling earthen chimneys at the entrance of their nests. Digger bees construct from one to several cells per nest, depending on the species.

Many species within the group are specialist pollinators of certain flowers. For example some *Melissodes* species specialize in thistles (*Cirsium*) or *Callirhoe*. Species of *Diadasia* are dedicated in their foraging habits, some visiting only sunflowers (*Helianthus*), others preferring cactus flowers (Cactaceae), and others flowers of the morning glory family (Convolvulaceae), mallows (Malvaceae), or evening primrose family (Onagraceae). The wonderfully named shaggy fuzzy-foot bee (*Anthophora pilipes villosula*) is a specialist pollinator of blueberries, although it is also useful for pollinating spring-flowering crops such as apples. It was introduced into the United States from Japan. The shaggy fuzzyfoot is especially desirable as a pollinator of crops because females work on rainy days, are fast flyers, and cover much territory when other bees fear to fly. Some *Anthophora* adorn the entrance to their nest with mud turrets (Figure 126).

Among the specialist pollinators within the digger bees must be listed the squash and gourd bees (*Peponapis, Xenoglossa*). These robust bees range in length from about $1/2$ to $3/4$ inch (12 to 18 mm). Although the body is basically black, the head and thorax are abundantly covered in yellowish hairs, and the abdomen has hair bands of similar coloration. Squashes, pumpkins, and gourds (*Cucurbita*) are New World endemics with both male and female flowers on the same plant, and they require a pollinator to move pollen from male to female flowers. Thus, in geological times, cucurbits would have required the services of a pollinator endemic to the area in which they grew. Squash and gourd bees coevolved to do this job in nature, but they were largely co-opted by honey bees as these cucurbits were hybridized and grown in agricultural monocultures. Here in the western United States where I live, there are still several native cucurbits that are pollinated by local squash bees. These bees arrive early in the morning, before the honey bees, and thus do the job they evolved to do.

Squash and gourd bees are the subject of a Squash Pollinators of the Americas Survey that was established by the U.S. Department of Agriculture's Pollinating Insect Research Unit in Logan, Utah, in an attempt to determine just how important they are. As stated at their website, if squash bees "prove to be ubiquitous, prevalent, abundant and effective, then this would be the first case for unmanaged, native nonsocial bees playing a key role for production of any agricultural crop." Thus, squash bees would seem to be potentially of great importance. Based upon the first year's survey this proved to be true: "It appears that an unmanaged group of nonsocial native bees—the specialist squash bees—are largely responsible for the production of cultivated squashes across North America, and by extrapolation, to much of the Americas" (Cane 2005).

Nomadine or Cuckoo Bees

The cuckoo, or cleptoparasitic, habit is found randomly dispersed in several families of bees (that is, Halictidae, Megachilidae, and Colletidae), but nomadine bees (Apidae, subfamily Nomadinae) exclusively exhibit this behavior. There are more than 1200 known species of nomadine bees in the world, by far the largest number of cleptoparasitoids of any bee family. Nearly half of all nomadines occur in North America. Although related to digger bees, they appear completely unlike their cousins. These are essentially hairless, wasp-like bees in varying shades of black, white, yellow, and rust (Figure 127). They range in size from about $1/8$ to $5/8$ inch (3 to 15 mm) and have given up their pollen-collecting ways to live as thieves. These bees are cleptoparasitoids because they enter the nests of other bees and highjack the provisions for their own offspring. This is basically accomplished in two ways. One is for the cuckoo to enter the host bee's nest as it is being provisioned and lay its own egg on the pollen provisions. The other is for the cuckoo to await the nest's completion, then chew its way into the host cell, lay an egg, then patch the cell up when it leaves. Species of one common genus of cuckoo bee (*Nomada*) are known to parasitize selected bees in all the other bee families, including its own, whereas the related genus *Triepeolus* attacks only a few families, and *Epeolus* attacks only members of the family Colletidae (yellow-faced bees).

Bumble Bees

We now move into the world of the truly social (that is, eusocial) bees. Worldwide there are more than 200 species of bumble bees (*Bombus*; Figure 128), with about 50 species in North America. These most often appear as colorful combinations of flying black-and-yellow fur coats, but there are a few black-and-orange or just plain black species as well. Bumble bees certainly receive much less press than honey bees, but unlike their cousins, most are true natives of our soils and have coevolved with our native plants. Bumble bees differ from honey bees in several respects. For one thing, neither bumble bee colonies nor the old queen survive the winter. In late summer or autumn a number of reproductives are produced, including males and new queens. Mating takes place outside the nest, either on the ground or in the air. Males die and new queens either return to the old abandoned nest or seek out an underground hibernation site in which to overwinter. The following spring each surviving female will become the founder of her own colony.

Bumble bees nest in a variety of situations, either above or below ground depending on the species. They are opportunists to a degree, seeking out hollow areas in which to build their relatively small nests. Abandoned rodent burrows are a preferred site, but I have seen nests in unused bird houses and have been told they use old mattresses and upholstered chairs lying in the dump. The queen begins her nest in the spring by constructing a hollow wax basin in which she places pollen and another in which she places nectar she has collected. When enough pollen has accumulated, she lays one to two dozen fertilized eggs and caps the basin with more wax. The collected nectar is used by the queen as fuel during this process and eventually for workers on their outbound flights. Then she sits astride the egg chamber, warming it with contractions from her wing muscles. Heating speeds egg hatch, which takes place in a few days. After the eggs hatch the queen takes off and gathers more pollen and nectar supplies, during which the larvae take one to two weeks to develop into adults. These then become the initial group of workers that create more cells, collect pollen and nectar, feed new offspring, and guard the nest while the queen continues to lay eggs. Workers perform functions as needed and may live from a couple of weeks to several months. Bumble bee colonies in temperate climates may reach a hundred or so individuals per nest. Heading into autumn the queen begins to lay unfertilized eggs, which develop into males, and workers supply more food to some of the larvae, which develop into queens. Reproductives leave the nest, mate, the males die, and the newly mated queens overwinter to begin a new colony the following spring. The old queen and her offspring perish.

Bumble bees are great pollinators of native plants, providing pollination to different varieties of plants than do honey bees, which are European natives. Because bumble bees are generally more active in cooler weather and lower light levels than honey bees, they are also adapted to service different sorts of flowers. Among edible plants, they are much better at pollinating tomatoes, peppers, squash, cucumbers, eggplants, and blueberries because these plants require buzz pollination (sonication) in which the bee dislodges pollen by vibrating the flower. Bumble bees are commercially available and managed as are honey bees, but they are more adapted for use in greenhouses to pollinate crops such as strawberries and tomatoes. *Bombus impatiens* is the only commercially significant native North American species, with more than 50,000 colonies sold per year (North American Pollinator Protection Campaign 2006). Attempts to introduce a European bumble bee, *Bombus terrestris*, into other regions of the world is meeting with controversy and opposition. This species has been introduced into Japan but now is declared an "invasive alien species" because of its negative effect on native bumble bees. In Australia, a continent with no native bumble bees, its introduction has been outlawed due to the fear that it may outcompete native bees (Aussie Bee, website).

About half a dozen bumble bee species resort to social (or brood) parasitism of other bumble bees. These are called cuckoo bumble bees and are often placed in their own genus (*Psithyrus*). Females of these species have no apparatus for collecting pollen, they cannot secrete wax, and they cannot create worker females. As a result they are unable to build and provision their own nests. To surmount these obstacles, they are well armored with thick skins and an excellent stinger, and they are sneaky. A cuckoo female invades the nest of an unsuspecting bumble bee species, and, if she is not killed outright, eventually assumes the odor of her new nest mates. Once accepted into the chemical fraternity that is a bee nest, her duty now is either to kill the rightful queen and control the workers or to slowly supplant the old queen until she fades out of the picture. The new queen will lay eggs of her own that will then be cared for by the workers of the former queen, and a new crop of social parasites will be born to strike yet another colony. This system works, in the end, because there are roughly eight times more social species to be usurped than cuckoo species to usurp them.

Now for some more bad news about bees. Bumble bees, like honey bees, are suffering declining numbers. Beginning in the 1990s researchers in the United States began to notice a decline in three common species of bumble bees. Their decline is likely due to a strain of fungus (*Nosema bombi*) introduced into the United States from

Europe. The decline is occurring elsewhere as well. In Great Britain there are between 15 and 20 species of bumble bee. According to a report by BBC News (February 27, 2008), at least three of these have become extinct and another eight are in decline. To understand this decline, some individual bees are being outfitted with tiny radio-frequency identification tags, which can be read by an electronic device placed at the colony entrance. By studying the comings and goings of individual bumble bees, scientists hope to learn more about their lives, not the least of which is why they are disappearing.

For those interested in bumble bee identification there is an interactive pictorial guide to species at the Discover Life website (see the list of useful websites). And for those without computers, there is an inexpensive little book entitled *The Natural History of Bumblebees* by Kearns and Thomson (2001) that features a photographic field guide to North American species. British species may be identified using the guides by Prys-Jones and Corbet (1991) and Edwards and Jenner (2009).

And Finally, a Word or Three About the Honey Bee

If I had to single out an insect that has received more attention than any other known to humankind, I'd pick the honey bee. A simple search for "honey bee" on the internet pulls up more than 4.5 million hits. Honey bees are not just the subject of popular and scientific books for both adults and juveniles, papers in scientific journals, and endless newspaper articles about killer bees or disappearing bees, but references to honey bees are made in movies, epigrams, poetry, ballads, songs, and mythologies. Then there are all the familiar products such as honey, wax, food, medical remedies, cosmetics, and drinks made from honey. Because this book can't possibly compete with that mountain of information and other bee species remain antithetically orphaned, I'm severely limiting my comments about the honey bee to three main areas: what a honey bee is, some basic honey bee facts that might prove useful at your next dinner party, and why they are disappearing.

We tend to think of "the honey bee" as a single species, without realizing that there are seven distinct species of honey bee, all of which originate in the Eastern Hemisphere. Within these species are

44 subspecies, that is, isolated geographical and often biological forms of the same species (Engel 1999). As gardeners we are most familiar with the European (or Western) honey bee (*Apis mellifera*), of which there are 28 subspecies.

In the Western Hemisphere, the honey bee we have known and loved for centuries is not native to our shores, but is an immigrant from Europe. Actually it's a bit more complicated than that, because it's at least two immigrants from Europe: the Italian subspecies (*Apis mellifera ligustica*) from southern Europe and the German (or Dark or Black) subspecies (*Apis mellifera mellifera*) from northern Europe. These forms were brought by the colonists in the early 1600s. Thomas Jefferson stated that Native Americans referred to the honey bee as "the white man's fly" because it preceded the advance of colonists across the continent (cited from Engel 1999). Several centuries later the Carniolan honey bee (*Apis mellifera carnica*) was introduced from the Balkans and the Caucasian honey bee (*Apis mellifera caucasica*) from the Caucasus. In recent times, things took a decidedly more confusing and malevolent turn. Enter the dreaded killer bee.

Sometime in the 1960s, The Americas faces yet another immigrant honey bee subspecies, one that created fear and trembling. This was the African subspecies (*Apis mellifera scutellata*) introduced into Brazil in the mid 1950s. Almost immediately the bee escaped confinement and, with a bulldog-like, overly aggressive tendency for self-protection, earned itself sensational headlines as the notorious killer bee. Truthfully, these bees simply objected to strangers anywhere near their home, and, unlike the Italian forms, held no reluctance to punish the transgressors. As the terror moved northward from Brazil and began interbreeding with its Italian, German, and Balkan cousins, it was rechristened the Africanized bee, which indicated a certain degree of civility (compared to its purely African form). The Africanized bee officially reached the southern border of the United States in the early 1990s, and by 2008 it had spread throughout much of the southern states from California to Florida.

There is not much more to say about Africanized bees, except to give several serious words of warning. If you live in a part of the country where these bees have become established, stay as far away from their nest as is humanly possible. Although individual bees flying about the garden pose little threat, when it comes to their headquarters they are merciless defenders. If a swarm takes up residence in or near your house, have a *professional* remove them. Second, a word of warning to hikers when walking in Africanized bee territory. Even at a distance the nests of these bees are surrounded by a very distinct buzzing sound as the hundreds of workers fly to and from its opening. The nest may be a tree hole high up in the forest canopy, or it may be a fallen log on the ground (I've seen both), but the sound of buzzing is a sure sign you are nearer to a nest than you want to be. A dignified, hasty retreat is in order. Walk away from the sound, not toward it. That is my advice. To my credit I sometimes listen to myself and have walked away from half a dozen encounters that might otherwise have proven less than pleasant.

Now to some basic facts about honey bees. Honey bees live in colonies of tens of thousands of sterile female workers ruled by pheromones produced by a single queen. All workers are daughters of the queen. In the wild, colonies may be housed in hollow trees, rock crevices, unused outbuildings, attics, or even exterior openings into stud walls. Very rarely honey bees build exposed nests in trees. When managed by humans, their homes are called hives (Figure 129).

The interior of a nest is made up of layers of vertical, double-sided honeycomb with six-sided cells. These combs are made from wax excreted by females from glands on the underside of their abdomen. Pollen is stored in some cells and honey in others. Honey is simply floral nectar that foraging bees bring back to the nest and regurgitate to other worker bees, who then allow water to evaporate. Entomologists, being the wits they are, often refer to honey as bee vomit.

The queen is an egg-laying machine, depositing one egg into each cell. She may lay as many as 2000 eggs per day and has a life span of five years if she's lucky. Each worker lives about six weeks and systematically rotates through specialized jobs: colony cleaner, larval nurse, wax maker and comb builder, unloader of pollen and nectar from incoming workers and honey maker, nest guard, and finally foraging bee.

Upon returning to the hive, foraging bees can impart information to other bees by dancing. Information includes the direction to fly, the distance, and abundance of the food source. This is communicated in the dance based on the pattern, speed, and degree of repetition. If you need more details—and they are well worth knowing—

almost any book or internet source on honey bees will go into splendid detail. Or, read *The Dancing Bee* by Karl von Frisch, which I read as a high school student and can still recommend highly.

When the colony becomes crowded, new queens and males (drones) are produced and the old queen leaves with a swarm of workers to begin a new colony. During swarming the workers are quite docile; even though there may be tons of bees flying about, they are not particularly aggressive.

The new queen temporarily leaves the colony, flies into the air, and mates sequentially with a number of males, whose genitalia are ripped out and die as a result. The new queen, once multiple-mated, returns to the old nest and lays eggs, never to mate again. She may eventually leave to swarm, but her mating days are limited to one per lifetime.

Honey bees have a double whammy when it comes to sex and self-defense. If a male successfully mates his organs are ripped out, and if a female successfully defends herself, her stinger is ripped out. In either case, death is the result. Consider these choices the next time you feel sorry for yourself.

As an interesting aside, in Japan there is a predatory wasp that is 2 inches (50 mm) in length called the Asian giant hornet (*Vespa mandarina*, Vespidae) that can kill an entire nest of introduced Italian honey bees. A single adult hornet leaves a chemical signal (pheromone) at a hive to attract several dozen of its own kin to attack the colony, kill off its adult population, and consume the larvae, pupae, and honey, which it feeds to its own larvae. As few as 30 adult hornets can kill a colony 1000 times bigger. But when the same hornet attempts to leave its chemical signal in a nest of the native Japanese honey bee (*Apis cerana japo*nica), the bees instantly surround the hornet, overheat it, and kill it thermally before the single founding wasp can signal its kin. If this isn't an indication that "native is better," I don't know what is.

Sometime around 2004 U.S. beekeepers noticed a sudden decline in their overwintering honey bee colonies. One-third or more of the nation's honey bees simply abandoned their hives, leaving their queen and immature siblings behind, never to return. The loss to beekeepers ran from 50 to 90 percent of their hives. In early 2007, the U.S. Department of Agriculture named the vanishing honey bee problem colony collapse disorder. According to various sources, however, colony disappearances have occurred for at least a century and have been called by different names including spring dwindling, fall collapse, autumn collapse, and disappearing disease. My favorite current appellation was coined by the British: Mary Celeste syndrome, named for an actual ghost ship mysteriously found deserted and under sail heading toward the Strait of Gibraltar in 1872. Leave it to the Brits to create a romantic sounding name for a perfectly mundane occurrence.

Although the mystery of colony collapse disorder has not yet been solved, it is not for want of endless attempts at possible explanations, such as: viruses, including Kashmir bee virus, acute bee paralysis virus, deformed wing virus, black queen cell virus, Israel acute paralysis virus, and several dozen more viruses, both known and unknown; fungi causing foulbrood; true parasites such as varroa mites (accidentally introduced to the United States in 1989 from Asia) and tracheal mites (introduced in the 1980s from Europe); insecticides (one study of 108 pollen samples revealed 46 pesticides, with as many as 17 different pesticides in a single sample); hive movement causing stress (yes, bees can be stressed); malnutrition due to drought, rain, and/or high temperatures that result in lack of bloom and thus lower nectar and pollen availability; air pollution because pollutants may block flower scent from spreading and thus bees can't find them; overwork due to high rates of pollination at certain times of the year; and all or some combination of the above.

Colony collapse disorder is not just some lame excuse to subsidize beekeepers, it is a serious problem for consumers as well. Recall that honey bee pollination is worth somewhere between roughly $7 billion and $16 billion to American agriculture. A loss of pollination by honey bees will affect much of our food supply. Even an ice-cream company has become involved with the disappearing bee. According to an article in the *San Francisco Chronicle* (Lochhead 2008) the folks at Häagen-Dazs have started a campaign (www.helpthehoney-bees.com) to halt honey bee decline. The reason is that more than 40 percent of its product's flavors are derived from fruits and nuts dependant on honey bees for pollination.

Without doubt the honey bee, whatever its form, is one of the most interesting of all insects in terms of social development and behavior. The discovery of its dance language and remarkable communicative abilities earned the German naturalist Karl von Frisch a Nobel Prize in physiology/medicine (1973) for his work with chemical communication among members of a honey bee colony. Von Frisch wrote a most enjoyable book called *The Dancing Bee*. Should anyone be interested in exploring the culture of bees from the historical aspect of human culture itself, then by all means read *Bee* by Claire Preston (2006), which only treats the honey bee, though it is not apparent from the title! A more nationalistic approach is provided in a book by Tammy Horn (2006), entitled *Bees in America*, which explores its subtitle of *How the Honey Bee Shaped a Nation*. All three book titles highlight the misappropriated umbrella term *bee* as a standard bearer for the honey bee. Such is the dominance of this single bee in our lives.

9

The Garden's Recyclers: Ants

THE CATEGORY ANT is unique among Hymenoptera because it represents but a single family, Formicidae, unlike the sawflies, parasitic wasps, predatory wasps, and bees, each of which is composed of many family groups. Still, what ants lack in family diversity they make up for in absolute quantity, weight, environmental importance, and notoriety. Unfortunately this does not translate into much help because what they make up for in quantity, weight, importance, and notoriety, they lack in characteristics that allow us to easily distinguish one species from the other. The main trouble with ants is, well, basically they all look like ants, with the possible exception of size and color. Technically, of course, and upon close inspection, ants are bristling with distinctive and often bizarre characters. Features such as number of antennal segments, relative lengths and numbers of mouth parts, number of teeth on the mandible, eye size, abdominal segmentation, and so on are used simply to identify ants to genus, never mind to species level. But in spite of many such characteristic differences, in practical terms most ants are difficult to identify. This difficulty is made more complicated by at least three factors. First, there are currently over 12,500 described species of ants known throughout the world (antbase.org, see the list of useful websites) and about 750 species in North America (Fisher and Cover 2007). Second, many ant species are polymorphic, that is, several forms (and intermediates) of diverse appearance occur in the same colony. Finally, different species of ants may occur in the same nest, compounding the problem of polymorphism within a single species and confusion between two different species inhabiting the same space.

Ants, as we are likely aware, occur nearly everywhere on Earth, including some extremely demanding places such as the Sahara Desert and the Australian outback. Several ants hold records for their ability to withstand more heat than any other insect. Australian species have survived laboratory temperatures hovering near 129°F (54°C) for an hour, but when actually foraging they don't even leave their nests until the ground temperature is 133°F (56°C) (Sherwood 1996). It is due in part to an ant's ability to cope with such environments that they are so successful—that, and the fact that they are social and occur in vast numbers no matter where they live.

According to Hölldobler and Wilson (1994), when combined, all ants in the world taken together weigh about as much as all humans. This statement is a bit old, so I'm not sure it would apply today with our current human obesity epidemic, but by all accounts ants are a principal component of habitats around the world. For example, approximately 2.5 acres (1 hectare) of Amazonian rainforest soil was reported to contain 8 million ants and 1 million termites, together accounting for three-quarters of the total weight of all insects in the forest. In assessing ant populations in Brazil, Hölldobler and Wilson (1990) stated that ants exceeded the weight of terrestrial vertebrates by four times. When we get to the subject of colonies, below, the numbers become even more staggering. Ants, it would seem, are not to be dismissed as some piffling, inconsequential irritant.

Despite their numbers, it is probably safe to say that many ants go unnoticed by the gardener, even though they are among the most numerous insects to be found, if not seen, in our gardens and our homes. And when I say "in our homes" I don't just mean hanging around the kitchen pantry, I mean in a home's structure. Even though ants are often considered a nuisance around and in the house, they seldom wreak excessive havoc in the garden except perhaps for some major regional exceptions such as the red imported fire ant, mound-building ants, harvester ants, leafcutter ants, and a few others that can cause extensive cosmetic (and sometimes painful) disruption to our gardens or ourselves. If these ants are ignored, the primary damage to our gardens lies in the disruption to plant roots, mounding of soil at plant crowns often leading to rot, mounding of soil in pathways or between pavers, the protection of plant-sucking insects such as aphids (Figure 130), and even the spread of plant pathogens such as bacteria and fungi. In spite of such negative actions, ants fulfill a major role in the environment by aerating and mixing the soil, enhancing water infiltration, recycling and incorporating dead and dying organic matter (both plant and animal) and nutrients, and, in many species, actively dispersing seeds of perhaps as many as 50 percent of the world's plant species.

My intent in this chapter is to give a generalized overview of an ant's life to better understand their potential effect on the garden or the gardener (Figure 131). I then discuss a few common ant species with regard to their diversity, nesting habits, and habitat choice. As with the bees, much literature is currently available on ants so that those interested will not suffer a dearth of reading material if they are so inclined. *The Ants* (Hölldobler and Wilson 1990), both a massively technical and Pulitzer Prize–winning tome, is the absolute bible on the subject, but at more than 700 pages and 7.5 pounds (3.4 kg) its authors later noted that it was "not a book one casually purchases and reads cover to cover." As an alternative and more literary book, they wrote *Journey to the Ants* (Hölldobler and Wilson 1994). *Ants of North America: A Guide to the Genera* (Fisher and Cover 2007) is geared toward the scientific identification of all 73 North American ant genera, with 180 color photographs (at least one of each genus) and 250 line drawings. The distribution and ecology is given for each genus as is a list of all North American species. The book might be a bit too technical for all but the most dedicated student of ants, but the images alone provide a greater respect for the variety and diversity of ants than any other resource I know of. For those who hanker toward the literary aspects of wildlife, a book simply called *Ant* (Sleigh 2003) will fit the bill. Several online data sources for all things "ant" are antbase.org, AntWeb, and *Notes from Underground* (see the list of useful websites). And finally, if you really, really want to learn about ants, there is an annual 10-day workshop called "Ant Course" usually (but not always) held at the American Museum of Natural History's Southwestern Research Station in southeastern Arizona (see the list of useful websites).

In addition to numerous technical books on the subject, ants have become subjects of legend. Within the past half century they could easily be mistaken as the emblem for our insipient cultural evils as well as a signal of environmental challenges to come. The movie "Them!" (1954), which featured giant, mutant ants created by atomic testing in the Arizona desert, was cited as being the first good movie (by contemporary standards) about giant insects (and spiders) that led to its rapid imitation by films about giant grasshoppers, tarantulas, praying mantids, and scorpions (Berenbaum 1996). Following upon the theme of atomic detonations came the film "Empire of the Ants" (1977), whose vision of environmental contamination was highlighted by toxic chemical wastes in which humans are enslaved by giant ants. This presupposes, of course, that we don't annihilate ourselves before ants get the chance. It would likely be the better option.

Ants often have been exemplars for industriousness and super-smart superorganisms, but one of America's great writers, Mark Twain, once wrote a most scathing assessment of ants as false illustrations of such notions. After watching ants stumbling all over themselves for hours, in *A Tramp Abroad* (1880) Twain penned a humorously scathing and lengthy diatribe against their so-called intelligence, concluding, "In the matter of intellect the ant must be a strangely overrated bird," and "I have not yet come across a living ant that seemed to have any more sense than a dead one." Writing later in *What Is Man* (1906), Twain appeared to have softened his position when he wrote, "As a thinker and planner the ant is the equal of any savage race of men; as a self-educated specialist in several arts she is the superior of any savage race of men; and in one or two high mental qualities she is above the reach of any man, savage or civilized!" In this conversation-based essay, Twain took the role of an elderly cynic ("Old Man"), so one is left to determine for themselves if his earlier opinion had changed, or—and this is more likely—his opinion of humans had been so inversely lowered as to view ants from a relatively more favorable position. Having by then arrived at the age of 71, perhaps he was simply playing the role of curmudgeon for his own amusement.

Basic Biology

Because ants comprise a single family of morphologically similar creatures, there are bound to be biological similarities among its species. For this reason we'll first examine the basic ground plan of ant biology before investigating some of the variants to be found in our gardens. Although the basics of ant life may appear relatively simple on the surface, the details can certainly prove challenging and complicated beyond belief.

Colonial Life

Of all Hymenoptera, ants are acknowledged to be the only exclusively social (that is, eusocial) group; no one has yet found an ant species that lives as a hermit. Of course, no one has seen every ant, either, so anything might still be possible. All ants live in colonies of their own making or as social parasites in colonies of other ants. Depending on the species, these colonies may reside in the soil, under rocks, in dead wood, in timbers, in hollow spaces of every kind from electrical boxes to acorns, in aboveground carton nests constructed of organic material, and even plant-produced domiciles. Colony size depends upon the particular species, with the smallest known colony having as few as 10 to 12 individuals and larger single nests ranging up to millions. I say "single nest" because some ant species form supercolonies in which groups of nests coalesce and extend for hundreds or even thousands of miles. These colonies contain huge numbers of individuals. In Japan one such supercolony is reported to extend across 675 acres (270 hectares) and contain more than 300 million workers and 1 million queens living in 45,000 interconnected nests (Hölldobler and Wilson 1994). Recent discoveries of nesting abilities for the Argentine ant (*Linepithema humile*) suggest that in the Mediterranean region interconnected, underground supercolonies extend over 3600 miles (5700 km) in length (Giraud et al. 2002). There is some debate about the genetic relatedness of ants throughout the colony, but whatever the outcome of that debate, it is unlikely that members in one area of a supercolony will ever encounter members in distant areas. (See additional comments below under "Argentine ant.")

A basic ant colony consists of three castes: a queen, worker females, and males. In some species there may be several discrete classes of female workers—called minor, median, and major—with each size differing somewhat morphologically from the other. These are polymorphic species. In some cases, major workers become the soldier ants of the colony with disproportionately enlarged heads. Fire ants (*Solenopsis*), carpenter ants (*Camponotus*), and leafcutter ants (*Atta*), for example, have polymorphic worker females, whereas Argentine ants (*Linepithema*) do not.

The basic life history of an ant is this. At some point in the year, a reproductive cycle is begun in which winged males and queens are produced (one exception is army ants, in which only males are winged). The reproductives emerge en masse from the nest, a veritable life-swarm, with mating taking place during flights. The flight soon ends with its participants landing on the ground, the males dying, and the females shedding their wings. In most species (except social parasites, see next section), these queen females begin a new nest, first laying eggs and producing a batch of sterile female workers. These workers begin the nest-building process while the queen continues to increase the female working force by laying eggs.

As colony size increases, workers first care for the newly produced eggs and developing larvae, then become nest excavators and cleaners, nest defenders, and finally foragers. Colony order is maintained by a combination of different chemical signals called pheromones, which are produced in various areas of an ant's body and detected by an ant's antennae. These chemicals are used to form scent trails for foraging and alarm signals in case of attack, for caste and member recognition within the colony, and to signal loss or decline of a queen. Pheromones are also passed along by the process of trophallaxis (Figure 132), in which individuals store food in their crop and regurgitate it to other colony members.

One occurrence common to any gardener is that a rock or board is moved and ants are seen frantically scurrying about with blobs of white material in their jaws. These are the eggs, larvae, and pupae of their sisters, which are not stored in cells constructed from materials, as is the case with bees or wasps, but occur in clusters that can be moved about when necessary. As with nearly all Hymenoptera, if the larvae develop from unfertilized eggs they become males and if from fertilized eggs they become females. Differentiation of female ants into queens or workers is determined by the quality of nutrition obtained by each larva. This means that queens can be created as needed.

Social Parasitism and Its Complications

In social bees and predatory wasps, some species have adopted the role of social (brood) parasite in which one species surreptitiously invades the home of another instead of constructing its own nest. In North America nearly 15 percent of all ant species exhibit some form of this behavior (Fisher and Cover 2007). Although social parasitism is found throughout the ant family, two-thirds of social parasites occur in just two genera of ants, *Formica* and *Lasius* (subfamily Formicinae). Myrmecologists (scientists who study ants) have managed to obfuscate the category of social parasitism further into the subcategories of temporary parasitism, slave-making, and permanent inquilinism. As if life were not complicated enough, ants, it seems, are at the apogee of the socialistic pyramid and thus require more explicit terminology to explain their tiny, if not complicated, lives.

In the case of temporary parasitism (that is, typical social parasites) a newly mated and wingless queen of one ant species infiltrates the colony of a different species. Ant recognition is based largely on chemical signals, and the invading queen is odorless, much as a newly pupated member of the colony, so that once the new queen gains the protective cover of the colony's odor she becomes an honorary member of that species. In some species queens may fight their way into a colony or be carried in by a worker. In both cases they risk detection until their own colony scent wears off and they gain that of the new colony. In these so-called temporary social parasites, the newly accepted queen usually kills the resident queen and then uses the host workers to raise her brood. In this case, the parasitism is temporary until such time as the invading queen and her progeny have totally wiped out the host colony. I'm never quite certain of my semantics, but this sounds rather more like a permanent parasite than a temporary one!

In the case of slave-making ants (technically referred to as dulotic ants), the invading ant usually enters the colony of a different ant species and carries the eggs, larvae, or pupae back to its own colony. These stolen progeny soon gain the odor of the new colony, and when they become adults they are treated as working members of the colony even though they are a totally different species. *Protomognathus americanus*, a common species of the eastern United States and

Canada, nests in old plant stems and insect galls, enslaving at least three species of another ant genus (*Temnothorax*). The latter genus itself contains a species that enslaves the same three species of its own genus as does the former! Some of these slave-making ants occasionally raid colonies of their own species, thus enslaving their fellow ants into forced labor.

In North America there are some dozen species of so-called permanent inquilines, which are sometimes referred to as workerless social parasites. In this case, the mated, invading queen ant of one species lives in the host colony of another, being fed by its workers to produce not her own bevy of workers—which would eventually overwhelm the colony as with temporary parasites—but reproductive offspring. Thus, the host colony keeps its queen and workers intact while at the same time diverting resources to increase the brood of the invading queen. Eventually these reproductive offspring mate, and the newly inseminated queens fly off to join yet another hapless, unsuspecting colony.

This overview of the lives of parasitic ants is, frankly, limited and simplified to the extreme. These behaviors become ever-more complicated the closer one looks at individual species of ants. If the simple basics seem overly complicated and confusing, then you are in good company. Ants, it would appear, are not quite the simple, mindless little creatures you thought they were.

Ant Associates

Because ants are so abundant, widespread, and successful in the environment, many other animals and even plants live with them (inquilines, parasitoids), imitate them (mimics), specialize in eating them (predators), or use them to aid in their own reproductive cycles (seed dispersal). In some cases these associations are given special names as a way to classify the interactions. Here are a few you might want to mention when the dinner conversation hits a lull.

Myrmecophiles are animals that specialize in living within ant nests. Theoretically *myrmecophiles* should refer to "ant lovers," but the term has become somewhat ambivalent in its usage and can refer to nearly half a dozen different subcategories that even include insects that kill ants, which to me is not necessarily a great show of love.

Figure 94. *Sphex jamaicense* (Sphecidae), a Caribbean species occurring in Florida, is a ground-nesting cricket hunter. Photo by Susan Ellis, Bugwood.org

Figure 95. Species of grass-carrying wasps (*Isodontia*, Sphecidae) nest in preexisting cavities, plugging the cells and the nest entrance with wadded up grass blades. Their usual prey is crickets and katydids. Photo by Bob Carlson

Figure 96. *Crabro monticola* (Crabronidae) is a ground-nesting wasp that provisions its cells with adult flies. Up to 60 different fly species have been reported as prey. Photo by Howard Ensign Evans, Colorado State University, Bugwood.org

Figure 97. Species of *Ectemnius* (Crabronidae) are ground or twig nesters, depending upon the species. Adult flies of many families are their prey. Photo by Bob Carlson

Figure 99. Psenine wasps (Crabronidae) commonly use planthoppers and their relatives to provision their nests, which may be in the ground or in twigs, depending upon the species. Photo by Bob Carlson

Figure 98. Female *Clypeadon* species (Crabronidae) hunt worker ants as provisions for their offspring. Females either lie in wait or provoke ants into attacking them. Once the ant is stung, it is carried back to the wasp's ground nest in a structure called an ant clamp located at the tip of the abdomen. Length 1/2 inch (12 mm).

Figure 100. Species of *Astata* (Crabronidae) are ground nesters that hunt stink bugs. Photo by Bob Carlson

Figure 101. A common cicada killer (*Sphecius speciosus*, Crabronidae) as she begins digging her nest. Photo by Carll Goodpasture

Figure 102. There are many minute predatory wasps, among which *Pulverro monticola* (Crabronidae; left) is one of the smallest. This wasp, less than 1/8 inch (3 mm) in length, captures thrips, which it carries in its jaws back to its underground nest.

Figure 103. This spider wasp (Pompilidae) was spotted dragging its paralyzed prey to some protected spot in which to secure it and lay an egg. Because the wasp faces backward as she drags her prey, the spider is dropped numerous times as the female tries to figure out where she needs to go.

Figure 104. Tarantula hawks (*Pepsis*, Pompilidae) are not just deadly to tarantulas, their sting is among the most painful of wasps. As a rule, any brightly colored wasp is advertising an ability to defend itself. Of course, males cannot sting (some can bite), but appearing the same as females provides protection from predators. Female pompilids generally have curled antennae, as does this one, whereas males' are straight. Before picking up a tarantula hawk male, you must have a high degree of confidence in your ability to assess "curl." That's why I've never picked one up.
Photo by Carll Goodpasture

Figure 105. Mud daubers (*Sceliphron cementarium*, Sphecidae) are attractive wasps not easily provoked to sting. Photo by Bob Carlson

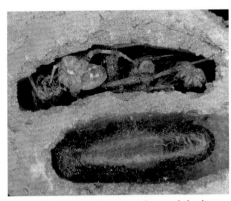

Figure 106. An inside view of a mud dauber (*Sceliphron cementarium*, Sphecidae) nest. The oblong cocoon is that of the wasp. The other cell is full of spiders and an unseen wasp larva. Mud dauber nests pose no threat to humans, as adult wasps do not live in them. Photo by Carll Goodpasture

Figure 107. Species of *Trypoxylon* and *Trypargilum* (Crabronidae) are spider hunters, many of which nest in preexisting crevices, hollow plant stems, abandoned beetle burrows, or even abandoned mud dauber's nests. Some, such as the *Trypargilum* shown, are known as pipe organ wasps because they construct large, multiple-tubed mud pipes attached to rocks, bridges, or buildings. Length 1/2 inch (12 mm).

Figure 108. Pollen wasps (*Pseudomasaris*, Vespidae, Masarinae) are predatory wasps that hunt pollen. It scarcely makes any sense, but there you have it. Length 5/8 inch (15 mm).

Figure 109. In the American Southwest, local paper wasps (*Mischocyttarus*, Vespidae) often appear confused as to where to build nests. Sometimes they are in the open, sometimes in enclosed spaces, and sometimes on light bulbs.

Figure 110. The nests of bald-faced hornets (*Dolichovespula maculata*, Vespidae) generally hang in trees where they are out of sight. When visible they can become the target of misguided individuals who throw rocks to see what happens. Given a choice, it's best not to see. Photo by Jerry A. Payne, U.S. Department of Agriculture, Agricultural Research Service, Bugwood.org

Figure 111. An excavated ground nest of the southern yellow jacket (*Vespula squamosa*, Vespidae). Disturbing any yellow jacket nest is risky, especially when the small entrance hole is no guide to the enormity that may lie below. Photo by Gerald J. Lenhard, Bugwood.org

Figure 112. A close-up view of a yellow jacket (*Vespula*, Vespidae). Some might say too close up. Photo by Carll Goodpasture

Figure 113. The European hornet (*Vespa crabro*, Vespidae), which is fond of killing honey bees, was introduced into the United States many years ago. These wasps appear to enjoy hanging upside down as they eat. Photo by Carll Goodpasture

Figure 114. Paper wasps, in this case *Polistes canadensis* (Vespidae), build paper nests commonly found attached to the eaves of homes. These wasps, along with their cousins *Mischocyttarus*, are the most docile of the social hunters, but you still don't want to pick them up.

Figure 115. *Mischocyttarus* species (Vespidae) are common paper wasps in the southwestern United States. They appear much like Polistes, but have a more slender abdominal petiole.

Figure 116. *Polistes dominula* (Vespidae) appears more like a yellow jacket than a paper wasp. This species was introduced from Europe and appears to nest just about anywhere, including mailboxes. Photo by Carll Goodpasture

Figure 118. The blueberry bee (*Osmia ribifloris*, Megachilidae) is a native species of the western United States, but can be used to pollinate blueberries. Photo by Jack Dykinga, U.S. Department of Agriculture, Agricultural Research Service, Bugwood.org

Figure 117. Leafcutter bees (*Megachile*, Megachilidae) gain their name directly from what they do. The bee (left) has just cut a circle from a *Campsomeris* leaf and is resting with it held between her jaws and legs. Right, the characteristic near-perfect circular cutouts left behind are commonly found on roses. These cutouts are used as is (not chewed) to line the cells of this solitary bee.

Figure 119. The wool carder bee (*Anthidium manicatum*, Megachilidae) appears more wasp than bee-like. The carder bees collect plant fibers to form their cottony cells. Photo by Bob Carlson

Figure 120. Some bees are cleptoparasitoids of other bees, invading their nests, killing the larval occupant, and replacing it with their own progeny. *Coelioxys* (Megachilidae) is such a bee. A male is shown here. Though the projections on his rear end appear to be multiple stingers, they are merely harmless extensions of the last abdominal segment. Photo by Bob Carlson

Figure 121. Mining bees (Andrenidae) are ground nesters, and species of *Perdita* (*P. arenaria* shown here) prefer sand. These bees are tiny, the smaller ones reaching about 1/16 inch (1.5 mm) in length.

Figure 122. Female yellow-faced bees (*Hylaeus*, Colletidae) carry pollen internally rather than on their legs. Mostly black and hairless, these bees are named for the yellow bands on their face. They are solitary, nesting in twigs or cavities. Length 1/4 inch (6 mm).

Figure 123. Halictid bees come in several colors, mostly dull. Little green bees (*Agapostemon virescens*, Halictidae) are attractive examples of what are commonly known as sweat bees. These are generally solitary, ground-nesting bees. Photo by Susan Ellis, Bugwood.org

Figure 124. Small carpenter bees (*Ceratina*, Apidae) appear nothing at all like their larger cousins, the large carpenter bees. These solitary bees nest in plant stems. Photo by Bob Carlson

Figure 125. Large carpenter bees (*Xylocopa*, Apidae) might be mistaken for bumble bees in flight. These solitary bees nest in solid wood. The males are both common and territorial, often challenging intruders to a stare-off. They are perfectly harmless.

Figure 126. Some bees nest in vertical banks. *Anthophora abrupta* (Apidae) is shown approaching its nest entrance, which has been modified into a mud turret. Photo by Scott Bauer, U.S. Department of Agriculture, Agricultural Research Service, Bugwood.org

Figure 127. Nomadine bees such as this *Tri-epeolus* (Apidae) do not make or provision their own nests but act as cleptoparasitoids in the nests of other bee species. Length 3/8 inch (9 mm).

Figure 128. Bumble bees (*Bombus*, Apidae) frequently nest belowground in abandoned rodent burrows. They construct globular wax cells in which to rear their young. Photo by Carll Goodpasture

Figure 129. A honey bee work camp. Honey bees are a commodity requiring frequent movement and colony disruption as well as long, hard working conditions in often inhospitable environments. There's no telling how much stress the average worker bee is under. Photo by Carll Goodpasture

256 •

Figure 130. Ants are commonly found tending plant-sucking aphids that excrete sugary substances (honeydew) resulting from the intake of phloem sap. Other insects that excrete honeydew include scales, mealybugs, treehoppers, psyllids, and caterpillars of several butterfly groups. Tending may take place above- or below ground, depending on the species. Ants protect these insects from harmful predators and parasitoids in exchange for the nutritious exudates. Photo by Carll Goodpasture

Figure 131. When viewed at close range, ants can be quite attractive in an odd way. Note the contrast between the shiny abdomen and the sculptured head of this species of New Mexican *Formica*. Photo by Carll Goodpasture

Figure 134. An army ant (*Nomamyrmex essenbeckii*) is heavily armored because it needs protection when attacking prey such as other ants, including the fearsome leafcutter ant (*Atta*). Length 1/2 inch (12 mm).

Figure 132. Two ants meet and exchange fluids in a process called trophalaxis, a nice term for regurgitating food. Not only does this distribute nutritional materials throughout the colony, but it also helps to maintain colony order through chemical signals (pheromones). The blackish background is composed of aphids, which have supplied honeydew that one of the ants is regurgitating. Photo by Carll Goodpasture

Figure 135. The common eastern black carpenter ant (*Camponotus pennsylvanicus*), well known to many homeowners. Photo by Clemson University, U.S. Department of Agriculture Cooperative Extension Slide Series, Bugwood.org

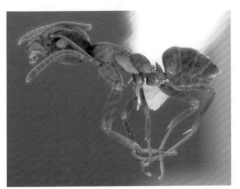

Figure 133. The Argentine ant (*Linepithema humile*) is among the world's top 100 invasive animal species. Length 1/8 inch (3 mm).

Figure 136. Not all carpenter ants are black, as attested to by this reddish carpenter ant (*Camponotus castaneus*). Photo by Bob Carlson

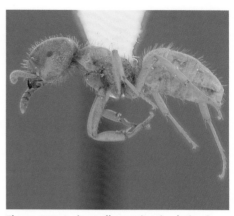

Figure 137. A citronella ant (*Lasius latipes*), so named for the citrus scent released when a colony is disturbed. Length 1/8 inch (3 mm).

Figure 138. Worker female of a Texas leaf-cutter ant (*Atta texana*) returning to the nest with a basic ingredient for the colony's fungus garden. Photo by Ronald F. Billings, Texas Forest Service, Bugwood.org

Figure 139. Face to face with a Texas leaf-cutter ant (*Atta texana*). Although they do not sting, their large jaws are as capable of pruning fingers as they are vegetation. Photo by Susan Ellis, Bugwood.org

Figure 140. The red imported fire ant (*Solenopsis invicta*) is an economic disaster and all-around nuisance in its introduced range. At the upper center is a fire ant decapitating fly (*Pseudacteon*, Phoridae), introduced into the United States in an attempt to control the ant. The fly must lay its egg on an ant's neck to be effective.
Photo by Scott Bauer, U.S. Department of Agriculture, Agricultural Research Service, Bugwood.org

Figure 141. The rough harvester ant (*Pogonomyrmex rugosus*), a species native to the southwestern United States, clears a distinctive, generally rocky area around its nest opening.

Figure 142. Although primarily seed feeders, the rough harvester ant (*Pogonomyrmex rugosus*) also takes insect prey, as this unlucky grasshopper discovered. To the center right of the photo, an ant holds what appears to be a brown seed in its jaws, but it is the legless remains of a butchered true bug.

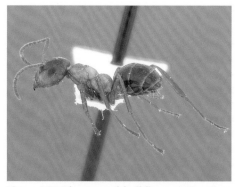

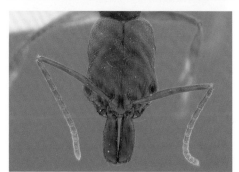

Figure 143. The mound-building ant, *Formica ulkei*, is somewhat unusual in that it builds its nests in prairies at the edges of swamps and marshes. Length ³/₈ inch (9 mm).

Figure 144. Trap-jaw ants (*Odontomachus*) are found in the southeastern United States, ranging south into the tropics. An adult worker of a Florida trap-jaw ant (*O. brunneus*; length ³/₈ inch, 9 mm, top) and a close-up of its elongated jaws.

For our purposes, myrmecophiles might best be considered those animals that are involved with ants in an obligatory but benign way (mutualistic) that is mostly beneficial to both parties. There are lots of things in an ant colony to attract such guests, including dead ants, live ants, fungi, plant roots, stored supplies, and other insect guests, so it should come as no surprise that an ant nest can be utilized by many different sorts of occupants besides the ants.

For example, ants may keep and tend underground colonies of root-feeding aphids, scales, or mealybugs for the honeydew they produce. In return, these sugarbags are protected from parasitoids and predators. Less well known, perhaps, is the tending of the larvae of lycaenid butterflies (blues, coppers, hairstreaks) by ants, some species of which they take belowground into their nest. The ants obtain sweet secretions from the butterfly larvae, and the larvae obtain protection in return, except for an occasional ichneumonid parasitoid or two that enters the nest and attacks them. Things may turn paradoxically sour for the ant, though, because after a while some species of lycaenid larvae turn into predators and eat the larvae of their ant hosts. Thus, not all myrmecophiles love their hosts except as a means to a personal end. There are some myrmecophiles that are merely scavengers. For instance, silverfish (Thysanura) eat detritus commonly found in ant nests. Numerous species of beetles (Coleoptera) act as scavengers but may also consume live ant larvae. Among these are included staphylinids (Staphylinidae), darkling beetles (Tenebrionidae), and scarabs (Scarabaeidae). Wingless, hunch-backed, ant-loving crickets (*Myrmecophila*, Orthoptera) appear to be primarily freeloaders in their host nests, feeding on regurgitated liquid they manage to elicit from a passing ant. Many species of mites (Acari) are ant lovers. There are even ant-loving lizards and snakes that live in ant colonies and eat other myrmecophiles such as beetles and sometimes protect the ants from attack from other ants (see leafcutter ants, below).

Myrmecophagy is the act of eating ants, and animals that eat ants are myrmecophagous. Some common specialist eaters of ants are anteaters (of course), antlions (Neuroptera), and horned lizards. Several groups of solitary, predatory wasps specifically attack ants as provisions for their own larvae. Some parasitic wasps attack ants, the most prominent of which is the entire family Eucharitidae, with more than 400 species that are specialist ant parasitoids. The larvae of these wasps gain surreptitious entry into the nest, where they are able to feed upon ant pupae. Because eucharitids are obligate ant feeders living in the ant nest, they may well be thought of as myrmecophiles. That is, they really do love ants . . . as dinner. But somehow this seems to miss the point, so I've included them here. The predatory larvae of flower flies (Syrphidae) also manage to find their way into ant nests to feed upon immature ants, whereas another family of flies (Phoridae) attacks adult ants aboveground. There are specialist ant-hunting spiders (*Zodarion*) that stalk their prey, attack it from behind, bite an ant's rear leg, and then wait for the ant to suffer paralysis. The hapless ant is then dragged away to a retreat for consumption (Cushing and Santangelo 2002).

Myrmecocleptophagy is a word I just invented because some entomological behavior seems not to fall into the above categories. Cleptoparasitism is the process of more or less surreptitiously stealing food from the mouth of another species. Myrmecocleptophagous animals are associated specifically with army ants (Ecitoninae). They act as camp-followers, eating or parasitizing insects flushed out as the ants march merrily along on their raiding sorties. Some members of the tropical antbird family (Thamnophilidae) specialize in picking off insect prey during such raids. Parasitic thick-headed flies (Conopidae) and tachinid flies (Tachinidae) lay their eggs in specific host insects fleeing before the marauding ant battalions. In some respects these could probably be called myrmecophilous insects except that don't actually live with the ants but take advantage of them when the opportunity arises.

Myrmecomorphy is ant mimicry, the situation in which a non-ant species appears to look like an ant. Myrmecomorphs include wasps (Hymenoptera), true bugs (Hemiptera), walking sticks (Phasmatodea), flies (Diptera), mantids (Mantodea), beetles (Coleoptera), thrips (Thysanoptera), springtails (Collembola), and mites (Acari). Spiders (Aranae) seem to be particularly prone to mimic ants because such species are found in at least a dozen different families. Obviously evolving into ant mimics has some survival value or these many different orders of insects and arachnids would not have ended up more like ants, say, than like spiders. The usual explanation is that an ant-like appearance presumably either confers protection from predators or offers a method of more easily approaching ants in order to kill and eat them.

Myrmecophytes are plants that specifically provide preformed cavities (domatia) in which ants can nest. This is a symbiotic relationship as the ant has a convenient living space and the plant receives protection from the ants, who ward off plant feeders. Additionally these ants are also known to prune adjacent vegetation as it encroaches on the home plant's territory. Myrmecophytes are sometimes called ant plants and are tropical in distribution.

Myrmecochory is the process of seed dispersal by ants. While moving organic matter from place to place, ants move seeds from near the parent plant to new ground, where they may develop without competition from their own parents. This behavior is often encouraged by the presence of nutritious structures (elaiosomes) on the seed. Perhaps 50 percent of herbaceous plants depend upon ants to assist in seed dispersal.

Some Ants of Note

Having given a short overview of ant biology and ant related-topics, let's look at some ants in more detail. Some of these ants are limited by geographic region and will be familiar mostly to gardeners in those regions. The red imported fire ant comes to mind for gardeners in the southern U. S., whereas mound-building ants may be more familiar to those in the eastern part of the country. Conversely, some species of carpenter ants are likely to be familiar to everyone, nearly everywhere. The ants below are listed alphabetically by common name so as to be easier to locate, but the genus, species, and subfamily is also given where possible.

Argentine Ant

The Argentine ant (*Linepithema humile*, Dolichoderinae; Figure 133) was introduced into the United States sometime in the late 1890s and was originally called the New Orleans ant, a name soon dropped due to local popular demand. This species is considered among the world's 100 top invasive animals. Presently it is found in California and parts of the southeastern United States. These ants form underground, interconnected supercolonies, which in California are believed to extend from San Diego to San Francisco (about 450 miles, 720 km). One such colony in Europe extends for 3600 miles (5700 km) along the coast from southern Italy to northern Spain. Members of these supercolonies cooperate with each other but attack when encountering genetically different members of their own species or other ant species. While Argentine ants do not sting, the primary problem for gardeners is that they obtain a great deal of their nourishment from plant-feeding and honeydew-producing insects such as scales, mealybugs, aphids, and psyllids. Because of this, Argentine ants fiercely protect these garden pests from predators and parasitoids, thus insuring the ever-increasing populations of those insects. According to a BBC report, "In California, they have displaced native ants, decreased the diversity of other native insects, affected the dispersal of seeds and even decreased lizard numbers" (BBC News 2004a).

Army Ants

Army ants, sometimes called legionary ants, comprise several genera and are largely tropical in distribution. They are included here only for curiosity sake, as they are rarely seen in the United States and certainly not in the vast hoards portrayed in documentaries. Several species do stray northward from the tropics into the southern and central United States. The general term *army ant* can refer to members of several different subfamilies, but the archetypal army ants of legend are all placed in the subfamily Ecitoninae. Army ants are predators of other insects, including ants. In the tropics huge colonies of army ants alternate between raiding (nomadic) and bivouac (or stationary) phases. Those species found in the United States are small (less than 1/4 inch, 6 mm) and are active mostly at night, so they are easily overlooked. Even so, some species share the alternate raiding/bivouacking behavior of their tropical cousins.

In the United States the main group of army ants belongs to the genus *Neivamyrmex*, which number some 25 species of the known 150. Although most of these species have limited distributions, the genus is represented throughout the southern half of the country, with several species extending as far north as Iowa (for maps of specific species distributions see *Online Catalog of Ants of North America*, http://www.cs.unc.edu/~hedlund/playpen/ants/GenusPages/Ecitoninae/Neivamyrmex.html). These ants are primarily predators of other ant species. *Labidus* is a genus of army ants represented by nearly 15 species in North and South America, but only *L. coecus*, reaches the United States, ranging from northern Argentina northward to Oklahoma, Texas, and Louisiana. The species lives underground as a predator of subterranean organisms so it is rarely seen wandering about. The genus *Nomamyrmex* is known from six species, with only *N. esenbeckii* (Figure 134) reaching into southern Texas, stretching northward from Costa Rica. It lives belowground but is a typical army ant in attacking its prey by group raiding during the day or night. This ant is built like a tank, wearing its skin like a heavy suit of armor—and for good reason. Leafcutter ants (*Atta*) are one of its main prey sources, which are attacked either on their foraging trails or in their colonies. Leafcutter ants are no slouch when it comes to returning aggression by way of their enlarged jaws, but apparently army ants always win the battle.

Carpenter Ants

Carpenter ants (*Camponotus*, Formicinae) are likely no stranger to those folks who own their home, or at least are under the impression they own their home. I discovered some years ago that the common eastern black carpenter ant (*Camponotus pennsylvanicus*; Figure 135) owned at least the wooden screened porch attached to the side of my brick house. Once that was torn down, they took immediate possession of a newly built addition placed over the site of the old porch. The reason for this was that caulking had not been properly done and wood under the newly installed patio doors became wet. Carpenter ants do not eat wood fibers, they merely tunnel through wood, and then primarily damp or moist wood. Once established they may also tunnel into sound wood. Numerous carpenter ants in the house, then, are good indicators that you've got more than simple ant problems—your house is rotting away as well.

Carpenter ant queens are the largest North American ant, reaching up to 1 inch (25 mm) in length, with workers up to ¹/₂ inch (12 mm). These are the largest ants found around and in the home. One positive aspect of carpenter ants is that they can't sting. Be forewarned, however, they can still gain your undivided attention by first biting and then spraying formic acid into the wound. Thus, carpenter ants cannot be handled with impunity. Individuals may be found wandering haplessly on floors and counters looking for any source of sweets or protein. The fact that you find them in the house does not necessarily mean they are living there; they may be living hundreds of feet away and merely using the cracks and crevices common to most buildings to gain entrance. A good gauge of probable infestation is quantity, as one ant does not a colony make. A colony of black carpenter ants may contain 15,000 or so individuals, so it's likely that a few dozen wandering around the house may indicate a problem. Baits designed for carpenter ants are a good way to disrupt and eliminate colonies living inside a home's structure, but they may take a few weeks to do the job.

There are some 1500 species of carpenter ants worldwide that constitute nearly 10 percent of all known ant species. Carpenter ants are mostly arboreal, living in standing and fallen dead tree trunks and branches. A few species are soil dwellers, though they might actually be living in buried tree trunks or even abandoned lumber (Figure 136). Only about 50 species of carpenter ants reside in North America, and only a handful of these pose household nuisances on a regional basis. The black carpenter ant (*Camponotus pennsylvanicus*) is all black and resides east of the Rocky Mountains. In the western United States and Canada, two carpenter ants create problems: *C. modoc* is black with reddish legs and *C. vicinus* is black with a red thorax. These two species have colonies with up to 100,000 workers. Not to be regionally outdone, the Southeast has the Florida carpenter ant (*C. floridanus*), in which the head and thorax are yellowish to red and the abdomen is black. Locally they are known as bull dog ants because of their strong bite. Colonies may contain up to 8000 individuals.

Citronella Ants/Yellow Ants

Citronella ants are so named because they emit a lemony odor when the nest is disturbed. They represent but a few species in the genus *Lasius* (Formicinae, Figure 137) of which there are nearly 60 species in North America and more than 100 worldwide. They are small, pale orange to yellow ants that live and feed underground. Citronella ants rarely come to the surface except during mating swarms and thus are rarely seen, though they may be extremely common underground. In my Maryland garden I frequently and inadvertently dug up nests of this ant without noticing it until I recognized the pleasant citrus scent wafting through the air. I would then see the ants scurrying madly about in their attempt to reclaim some semblance of normalcy. Citronella ants tend root-feeding mealybugs and aphids that secrete honeydew, which the ants then consume. In exchange for sweets, citronella ants protect these insects from predators. They seem to be of little consequence to the garden and can be a pleasant surprise for the gardener when encountering the unexpected fragrance as an accident of digging.

Leafcutter and Fungus Ants

Leafcutter and fungus ants (Myrmicinae) might better be referred to as gardener ants because they cultivate fungus in underground nests. All are members of the tribe Attini, which contains about 10 genera confined to the New World, mostly the tropics. Although all grow fungus, the classification of these ants is a bit mysterious, with the fungus growers and the leafcutters somehow differentiated in the minds of myrmecologists. All told, there are more than 300 species of attines, but only 18 occur in the United States and none in Canada.

Three species of leafcutters are known from the United States. *Atta texana* (Figures 138, 139) is a forest dweller found in Texas and Louisiana, whereas two desert dwellers include *Atta mexicana*, found in Arizona, and *Acromyrmex versicolor*, ranging from Texas to southern California. These species all range southward into Mexico. They are relatively large, reddish ants with workers ranging up to 1/2 inch (12 mm) in length. Their powerful jaws are designed to prune vegetative material from deciduous and evergreen trees, shrubs, and herbaceous perennials. Because large colonies can contain up to 2 million workers, leafcutter ants can cause extensive damage in areas where they are not wanted—gardens, for instance.

Leafcutter ants do not eat the vegetation, but chew it into small fragments that are arranged in protected chambers in which a single species of fungus is allowed to grow. From these fungal gardens, bits are eaten by the ants or fed to their larvae. Spores or infestations that arise from foreign fungal species are removed. Because the fungus grown in these gardens is a single species, as with all monocultures it is liable to sudden attack and potential collapse from harmful organisms. Being diligent farmers, leafcutter ants use several methods to defend against such potential disaster. They remove infected parts of the fungus garden to their dumping grounds, consume pathogen spores where the infection has not yet taken hold (BBC News 2004b), and harbor a bacterium that has antimicrobial properties.

Underground colonies range in size from several square feet up to 1 acre (0.4 hectare), with correspondingly increasing numbers of workers. Sandy or loamy soil is the primary requirement for establishing a nest, so leafcutter ants may be found wherever suitable soil occurs, including open fields, brushy areas, or forests. Depending upon the season, leafcutter ants work during the day (cool seasons) or at night (warm seasons).

Although leafcutters do not sting, their large jaws are as capable of pruning fingers as they are vegetation. In spite of these jaws, leafcutter ants are sometimes attacked by army ants, which generally win the battle. In some cases, however, leafcutters have an additional tool in their defensive arsenal. One leafcutter species in Trinidad has a myrmecophilous lizard that protects it from army ant invaders. This lizard may consume a few leafcutters from time to time, but the loss is apparently worth the protection provided. *Atta mexicana* also has an ally, in this case the Mexican short-tailed snake, which lives in its colonies and eats beetle larvae that are scavengers or predators of immature ants (San Juan 2005).

Fungus ants in the United States consist of some 15 species in three genera (*Mycetosoritis, Cyphomyrmex, Trachymyrmex*) with all but one species ranging from the southeastern to southwestern United States. Fungus ants differ from leafcutter ants in having smaller nests—a few hundred workers at most—and usually a single chamber for growing fungus. These ants culture fungus on bits of grass, insect droppings, dead insects, and leaf litter. Because of the small colony size and choice of substrate upon which to grow their gardens, fungus ants are less destructive than their cousins. *Trachymyrmex septentrionalis* is the only species extending northward from Texas to the New Jersey pine barrens, and it tends to act more like a true leaf-cutter in its harvesting of vegetation from larger trees and shrubs.

Fire Ants and Thief Ants

The genus *Solenopsis* (Myrmicinae) is represented in North America by about 40 species, but only one is usually thought of as "the fire ant," *Solenopsis invicta*, the red imported fire ant. Actually, there are six species of fire ants in North America, of which four are native and two are introduced. All are nasty biters and stingers, but the four native species likely would not be the subject of much concern were it not for the notorious red imported fire ant, which was accidentally introduced from Brazil or Argentina into the United States sometime prior to the 1930s. It was first recognized as unusual and reported to authorities in the early 1940s by a 13-year-old boy named Edward O. Wilson, now the world's foremost authority on ants and one of biology's most prodigious thinkers and prestigious writers. (Wilson, by the way, now an 80-year-old retired Harvard entomologist, is showing no signs of slowing down at this writing. He recently coauthored another book, *The Superorganism*, about termites, wasps, bees, and ants [Hölldobler and Wilson 2009], he's working on his first novel, and he's preparing a treatise that champions the notion that natural selection operates not just on genes but at many levels, including that of social groups.)

The red imported fire ant is now found throughout the southern half of the United States and occasionally strays or is transported northward, most often in nursery containers. At present its range seems to be restricted by cold climates, but that may eventually change by way of genetic mutations, warming climates, or a combination of both. One of the first signs of *Solenopsis invicta* is that the ants extrude a mound of soil (unlike harvester ants) at the nest entrance that is both unsightly to the tidy gardener, detrimental to plants, and an obstacle to lawn mowing. These mounds are especially harmful to agricultural lands because they create a physical hindrance to plowing, plant growth, and efficient harvesting. The second sign of *S. invicta* is that their mounded nests are easily disturbed, which results in a mass of angry ants immediately swarming out to ward off the intruder. In this respect they are much like Africanized bees. The third characteristic of *S. invicta* is that individuals first bite, then sting, leaving susceptible victims with terrible pustulating welts and sometimes serious medical conditions. This behavior is aggravated by the fact that large numbers of red imported fire ants will often first swarm over a victim unnoticed, then all bite and sting as if on cue. Many years ago I was wandering around a forest in Florida, head turned upward looking for cynipid galls, when I unknowingly stopped and planted my feet on the mound of a fire ant nest. I looked down and was amazed to see five or six dozen ants all biting and stinging at once. Fortunately I was wearing shoes made of rough suede, and the ants were mistakenly wailing away on the feet of what they perceived to be some dumb intruder when they were actually only attacking the shoes of some dumb intruder.

For years, attempts at controlling *Solenopsis invicta* have been largely chemical. In its homeland, however, the red imported fire ant is rarely a problem because it is kept in check by its natural enemies. One of those, a tiny, sinister fly, has been imported into the United States in an attempt at biological control of the ant. This is a phorid fly (Phoridae), which has the unusual habit of laying its egg behind the head of an ant—a superb example of aerial finesse (Figure 140). The egg quickly hatches, and the developing phorid initially feeds within the thorax, eventually moving to the ant's head, at which point the head falls off with the fly larva safely housed within. An excellent website to obtain many answers to questions about the biological control of *S. invicta* and basic information about the species is maintained by Larry Gilbert, director of the Brackenridge Field Laboratory at The University of Texas at Austin (see list of useful websites).

The red imported fire ant was not the first exotic fire ant to reach the United States, an honor that goes to the less well known black imported fire ant (*Solenopsis richteri*), also from Argentina. Although this ant arrived first, it has been superseded nearly out of existence by its cousin *S. invicta*. The remaining four species grouped as fire ants, but all native to North America, are found throughout the southern United States and pose few problems compared to the red imported fire ant, though it's best not to aggravate any of them, given the choice.

Excluding the fire ants, the three dozen remaining species of *Solenopsis* are called thief ants and pose no problem to human existence at all. These are tiny ants living in proximity to larger ants. Thief ants periodically raid their larger neighbors, stealing brood and food stores for their own sustenance.

Harvester Ants

There are approximately 25 species of harvester ants (*Pogonomyrmex*, Myrmicinae) in the United States with all but one occurring west of the Mississippi River. No species are known from Canada, but another 70 species occur in Mexico southward to southern South America. These ants are in the same subfamily as fire ants and can administer quite a nasty sting when provoked. I discovered this years ago when I was collecting them with my fingers under the mistaken notion that they didn't sting. They do, and painfully so. I no longer collect ants with my fingers.

Harvester ant nests are usually easily recognized by the virtual absence of living plants within yards (meters) of the normally single entrance hole. In some species the entrance is flush with the ground (for example, the rough harvester ant, *Pogonomyrmex rugosus*; Figure 141), but in others it is centered within a distinctive raised soil mound (such as the western harvester ant, *P. occidentalis*) or even a fan-shaped crater (the California harvester ant, *P. californicus*). The rough harvester ant, common in the American Southwest, surrounds its entrance with tens of thousands of tiny pebbles and sand grains that have been brought up from within the nest. These stone collars are somewhat distinctive, but they are so evenly spread out that they do not form a distinct mound. Some harvester ants clear a trail through the vegetation, such as grasses, over which they scout for food. These trails may extend hundreds of feet from the nest entrance. In terms of size, harvester ant colonies can be quite large, extending into the soil as much as 10 feet (3 m) and containing up to 10,000 workers.

Harvester ants are primarily seed collectors, but they are not adverse to slaughtering the odd insect prey or two (Figure 142). They are known to reduce rodent numbers by competing with them for seeds. Here in the Southwest, where I live, harvester ants are one of the common food items of the 13 species of horned lizards (or horny toads). From 50 to 90 percent of a horned lizard's diet consists of ants, and harvester ants make up a large part of that diet.

Should the topic of insect venoms or ant farms ever come up during polite conversation, you might want to impress your guests with several facts. You could mention that the Maricopa harvester ant (*Pogonomyrmex maricopa*), found only in Arizona, is considered to have the most toxic venom of any insect in the world (Meyer 1996). Alternatively you could mention that several species of harvester ants, usually the western (*P. occidentalis*) and the red (*P. barbatus*), are the most common ants supplied with children's ant farms. Or to particularly annoying guests, you might suggest that the next ant farm they buy be stocked with Maricopa harvester ants.

Honey Ants

Although a few honey ants (or honeypot ants; *Myrmecocystus*, Formicinae) occur in several regions of the world, the majority, some 30 species, reside in the desert regions of the southwestern United States and Mexico. Honey ants nest in the ground, and working foragers may be active at night or during the day, depending on the species. Although honey ants act as both scavengers and predators, with termites being one of their favored foods, they depend largely on plant exudates such as sap and nectar, as well as secretions from aphids and mealybugs. The young workers (repletes) are able to distend their abdomens in which is stored regurgitated excess nectar and honeydew. These repletes, nothing more than living storage vessels, hang like flagons from the chamber roofs until their contents are needed by the colony.

Mound-Building Ants

Mound-building is a behavioral category found in many ant subfamilies and genera throughout much of the world. In North America the group is represented by three species (the *exsecta* group) in the much larger genus *Formica* (Formicinae). Some other species of *Formica* (for example, the *rufa* group) are referred to as mound builders, but these are relatively uncommon and go by the name of thatching ants, because their mounds are interlaced with pine needles or vegetative debris. Mound-building ants are omnivorous, feeding on any available source of protein (such as other insects, dead or alive) and carbohydrates (such as honeydew). Their distinctive mounds, composed mostly of excavated soil and some organic debris, can reach several to many feet in diameter and height aboveground. Much as with harvester ants, mound-building ants clear an area around the nest. They do this by initially biting a plant's stem, then spraying formic acid onto the tissue. (Note to gardeners: Although they cannot sting, mound-building ants use this technique equally well on unwary fingers.) These ants can kill trees, shrubs, herbaceous vegetation, and grasses. Unlike most ants, mound-building ants may have more than one queen per colony. Most likely the gardener will not allow a mound builder to gain a foothold, but if left alone, these ants can become a serious nuisance in the garden.

Two species of mound builders are of little concern. The western mound-building ant (*Formica opaciventris*) is limited to the Rocky Mountains region, whereas *F. ulkei* (Figure 143) occurs in prairie regions of the upper Midwest extending into Canada. Of major concern is the Allegheny mound-building ant, *F. exsectoides*. It is the most commonly encountered North American mound builder, reaching from Georgia northward in the eastern United States and into southern Canada and westward to Colorado. This ant ranges in size from 1/8 to 1/4 inch (3 to 6 mm) and typically has a red head and thorax and a black abdomen. According to various literature sources, the enormity of this ant's work is not to be believed. For one colony, an early published work placed the number of worker ants at 237,000, ruled by a total of 1400 queens (Cory and Haviland 1938). Later data placed the number of workers at roughly 300,000 (Hölldobler and Wilson 1990). Although mounds are usually a few feet in diameter, they have been reported at 24 feet (7.3 m) across (Anderson 1999), though these may actually be a series of nests that have budded off from other colonies, much like the supercolonies in the Argentine ant.

Pharaoh Ant

The pharaoh ant (*Monomorium pharaonis*, Myrmicinae) is most likely a native of Africa that has spread all over the world. Although not a problem in the garden, it adapts well (too well, actually) to human buildings, where it relishes the warmth. This ant is so small (¹/₁₆ inch, 1.5 mm) and pale as to be almost invisible, but anyone who has them in the house knows they are a damned nuisance despite their size. Pharaoh ants are attracted to water, sweets, fatty foods, and dead as well as live insects. Although these ants neither bite nor sting, they unfortunately can invade human tissue given the chance, such as in hospital wards or nursing homes. Colonies range from a few dozen to several thousand individuals, but there may be many queens (up to 200). The colonies have an ability to rapidly expand (bud off) to form supercolonies or contract according to environmental conditions. Due to their omnivorous habits, their abilities to occupy any given space no matter how small and to withstand wide population fluctuations, and their association with human habitats, pharaoh ants are extremely undesirable as well as difficult to get rid of. Baits are about the only way.

Trap-Jaw Ants

In North America five species of trap-jaw ants (*Odontomachus*, Ponerinae, Figure 144) are reported, of which three occur in the southeastern United States. There are about 65 species distributed worldwide. According to a study at the University of California, Berkeley, a tropical species of this group has the fastest predatory appendages of any known animal (Patek et al. 2006). In plain language, this translates to ballistic jaws. It takes a tenth of a millisecond (145 miles per hour) for this ant to clap its jaws shut. Hölldobler and Wilson (1994) claimed it is "the fastest of any anatomical structure ever recorded in the animal kingdom." These jaws do double-duty because the ant can slap them against a hard surface, propelling itself up to 3 inches (75 mm) into the air to avoid its own predators.

Epilogue

AND SO WE REACH the end of our cursory yet curious glimpse into the contradictory world of Hymenoptera—a world that if it did not actually exist in our garden would be as improbable as any described by Jules Verne or as impossible as any painted by Salvador Dali. This book has been an attempt to distill my love of bees, wasps, and ants to as many readers as possible, readers who hopefully will be amused and amazed by a world that surrounds them, yet one that remains shrouded in obscurity.

In the very early stages of my scholarly insect studies I thought that parasitic flies might offer an interesting arena in which to spend my allotted time on this planet. I am certain they are every bit as fascinating as are parasitic wasps, but receiving little encouragement to study them, I searched for another aspect of entomology that might prove of interest. I gave the honey bees a try as a result of reading *The Dancing Bee*, this time receiving much encouragement to do so, but in the end it seemed that spending an entire lifetime attached to one insect might become a bit dull—interesting though honey bees may be.

It was my luck, however, to fall victim to one of the leaders in the realm of predatory wasps, Richard M. Bohart, who convinced me to work on his side of the Hymenoptera as an assistant. For several years I did so, but he did not realize that the little wasps, the parasitoids—the ones no one wanted to study—were invading my mind and would soon lead me on to the smaller, better things in life. In spite of becoming a traitor to his cause, the old man supported me in mine, and we remained colleagues and friends until his death in 2007 at the age of 93. I think he was actually delighted when any of his students broke rank and ventured into the unknown—as long as it was a hymenopterous unknown. As one of his former students remarked, "Bohart's first love was wasps, and his second love was students." I'd like to think that as one of his students I have channeled his first love in a manner that would have received his approval.

As to the outcome of my attempts, I will deem it a minor success if the next time you see an ant you think to yourself "Those hymenopterans are really interesting," just before you smack it into oblivion. It will be a mild success if you see a tiny speck and think to yourself "That could be a parasitic wasp," and let it fulfill its ordained destiny. And it will be an absolutely smashing success if you see a yellow jacket nest and think to yourself "Those wasps are actually beneficial, so I'll let them be, at least for now." I could scarcely ask for more.

Hymenopteran Families and Their Primary Larval Feeding Habits

AS WITH ANY ATTEMPT to consolidate information on a huge number of species, a simple starting point presents one problem and where to end presents an even greater challenge. There is no single up-to-date source from which to derive all information pertaining to Hymenoptera, and so it must be stitched together in bits and pieces from various sources. Too often these sources present incomparable information, or they do not even agree with one another. Biology is a lot like that.

The first table presented here is based largely on information limited to North America north of Mexico, which refers roughly to what zoogeographers call the Nearctic region. This is an artificial contrivance, but once south of the border all sorts of tropical elements become involved and data on these areas are sadly lacking in the literature. This region contains 75 families, all of which are also found in Europe except Anaxyelidae, Pelecinidae, Rhopalosomatidae, Sierolomorphidae, and Tanaostigmatidae (Fauna Europaea, see list of websites). The second table represents all remaining world families of Hymenoptera not reported from North America north of Mexico. These families represent fewer than 175 species, the life histories of which are largely unknown. It is likely that such low numbers are more a result of our miserable knowledge of the faunas of areas such as South America, Africa, Australia, and Asia than an indication of what we actually know. Neither table includes those families known only from fossils.

In these tables I attempt to present a coherent summary of what is known about Hymenoptera, but this requires a certain degree of explanation as to how and from whence it was derived. The explanation, itself, is nearly as complicated as the end result, but I give one so that users who doubt my cognitive abilities can redo the entire work for themselves if they wish.

Superfamily and Family Names

The families noted in these tables are based on the excellent summary paper by M. J. Sharkey (2007) entitled, "Phylogeny and classification of Hymenoptera." These families represent nearly 150,000 species. The order Hymenoptera is undergoing tremendous study by researchers all over the world and as a result family names may change without warning. Indeed, the entire classification may change in the near future. In both tables I provide the currently hypothesized evolutionary order of superfamilies from most primitive to most advanced; within each superfamily, families and subfamilies are listed alphabetically. Categories such as suborder and common name (for example, bees) have been given to help group and compartmentalize the superfamilies of Hymenoptera as seen by professional hymenopterists. As these may end up simply confusing the casual reader, I suggest ignoring everything but the family names unless you are a compulsive nomenclaturist. If a family has a common name, I add it; for example, spider wasps for Pompilidae or velvet ants for Mutillidae. Most families are simply called by a shortened name, such as scoliids for Scoliidae or pergids for Pergidae.

Subfamilies

In some cases I have listed subfamilies because these categories provided more precise information about feeding types and biological diversity found within a family. Some of the larger families are so massive that to break them down by subfamily would require endless pages of largely repetitive information. For example, Braconidae (29 subfamilies), Ichneumonidae (35 subfamilies), and Pteromalidae (39 subfamilies) are composed of a large number of subfamilies that do many things in many ways, and for these families I simply give a general overview. In many cases researchers are not entirely certain how to define the limits of the subfamilies to anyone's satisfaction, nor are the names likely to stand the test of time. To analyze them in great detail seems a bit pointless.

Numbers of World Species

I have attempted to indicate the number of world species for each family—not an easy task. General information in the following tables is compiled largely from *Hymenoptera of the World: An Identification Guide to Families* (Goulet and Huber 1993), which, although nomenclaturally dated at the family level, still contains reliable biological and numerical data if used with some degree of caution. More recent sources of information concerning numbers of species and biological attributes of each family have been used as well, but most are from obscure technical papers or websites. These sources are cited by author and date or website for those who wish to know the origin of such data. The complete source may be found either in the references, if cited by author, or the list of useful websites if cited as websites. In several cases unpublished information has been contributed directly to me by cooperative scientists and is cited as "personal communication." I have acknowledged these workers more fully in the preface.

The total number of world species for a family is representative of the known world fauna. Because all these numbers change yearly, almost none will ever be totally accurate. They are given merely to indicate the abundance of one family relative to another. An even more essential factor when considering numbers is that in many cases, especially the parasitic wasps, the number of suspected undescribed species actually outnumbers the described ones, so numbers given in the table may represent only half or less the total species we will eventually recognize. So poor is our knowledge of Hymenoptera that perhaps only one-tenth of the world's species have been discovered and named.

Numbers of North American Species North of Mexico

Unfortunately there is no source that lists all North American species of Hymenoptera, but there is one that lists all species north of Mexico until 1979. This is Nomina Insecta Nearctica (www.nearctica.

com/nomina/wasps/hymenop.htm), derived from *Catalog of Hymenoptera of America North of Mexico* (Krombein et al. 1979). I have used the web version to determine numbers used in the first table if I could not find them elsewhere.

Feeding Types and Hosts

Feeding type pertains to how the *larval stage* feeds. Adult Hymenoptera are essentially nectar feeders. Feeding patterns can be technically helpful in determining what a hymenopteran larva might be, especially when reared or found in association with host remains. The tremendous diversity of hymenopteran hosts and how they are attacked are also listed. The following biological terms are discussed further in chapter 3.

Predator: Generally, a predator eats its own catch, but a predatory wasp paralyzes an insect or spider prey, moves it to a concealed space, and lays an egg on or near it. It is the wasp larva that does the eating.

Inquiline: An insect that lives within the nest of another insect, without harming it, consuming surplus materials produced by the host.

Ectoparasitoid: A parasitoid wasp paralyzes a host and lays an egg(s) on it. The wasp larva feeds externally on its host.

Ectoparasitoid/predator: A parasitoid wasp places her egg(s) as would a typical ectoparasitoid, but after the egg hatches and the larva eats its host, it wanders off and eats other nearby hosts, thus acting as a predator.

Endoparasitoid: A parasitoid wasp injects her egg(s) into a host and the larva feeds internally.

Hyperparasitoid: A parasitoid that attacks another parasitoid in or on its host, but not the actual host itself.

Parasitoid sac: A specialized habit in which the wasp larva feeds inside the host, but part of its body is extruded from the host in a shell-like sac.

Cleptoparasitoid: A form of parasitism in which the larva usurps the food provided for the host (such as pollen in a bee cell), thus killing it. Sometimes a cleptoparasitoid kills its host outright and then eats the provisions.

Social parasite: In eusocial insects (some bees, predatory wasps, and ants), the queen of one species invades the colony of another, killing its queen; the colony workers then raise larvae of the invading queen and not their own.

Solitary: A female lays eggs where she wants her young to feed, then leaves. There is no maternal care, and females do not socialize with others of the same species.

Gregarious nesting: In some solitary bees and wasps, females nest near each other to the point where large aggregations of nests become apparent. Each nest is still worked independently by the female.

Communal nesting: In some solitary bees and wasps, females use the same entrance to a nest, in the soil or wood, for example, but each works independently within the site. There is no cooperation.

Eusocial: A few Hymenoptera maintain watch over their young, feed them progressively, have overlapping generations, and are ruled by a queen.

Hymenopteran Families Found North of Mexico

FAMILIES IN NORTH AMERICA NORTH OF MEXICO	NUMBER OF WORLD SPECIES	NUMBER OF SPECIES NORTH OF MEXICO	LARVAL FEEDING TYPE	HOSTS
SAWFLIES (Suborder SYMPHYTA)[1]				
Superfamily: Xyeloidea				
XYELIDAE	153	24	plant feeding/internal or external	pollen cones of pine (internal); buds and shoots of fir (internal); external leaf feeders of deciduous trees
Superfamily: Pamphilioidea				
PAMPHILIIDAE webspinning sawflies	301	75	plant feeding/external	leaf rollers, web spinners; woody plants including conifers
Superfamily: Tenthredinoidea				
ARGIDAE	800	70	plant feeding/external; one species mines leaves	leaves of many plants; leaf miner on *Portulaca*
CIMBICIDAE	196	12	plant feeding/external	leaves of shrubs and trees
DIPRIONIDAE conifer sawflies	140	45	plant feeding/external	conifer needles, cones
PERGIDAE	434[2]	4	plant feeding/external	leaf feeders on oak and walnut in North America; many plants elsewhere
TENTHREDINIDAE (seven subfamilies) common sawflies	5511	800	plant feeding/external or internal	primarily leaf feeders; miners or gallers of many different plants; rarely borers
Superfamily: Cephoidea				
CEPHIDAE stem sawflies	201	13	plant feeding/internal	grasses and woody plant stems
Superfamily: Siricoidea				
ANAXYELIDAE cedar wood wasp	34	1	plant feeding/internal	borers in burnt conifer trunks and branches
SIRICIDAE horntails	124	19	plant feeding/internal	in wood of conifers or hardwoods; females inject fungus into borings

FAMILIES IN NORTH AMERICA NORTH OF MEXICO	NUMBER OF WORLD SPECIES	NUMBER OF SPECIES NORTH OF MEXICO	LARVAL FEEDING TYPE	HOSTS
Superfamily: Xiphydrioidea				
XIPHYDRIIDAE wood wasps	138	10	plant feeding/internal	borers in trunks and branches of deciduous trees; females inject fungus into borings
Superfamily: Orussoidea				
ORUSSIDAE parasitic wood wasps	85	9	ectoparasitoid	wood-boring beetles and wasp larvae
PARASITOID WASPS (Suborder APOCRITA: PARASITICA)				
Superfamily: Stephanoidea				
STEPHANIDAE	100	6	parasitoid	wood-boring beetle and wasp (horntail) larvae
Superfamily: Trigonalyoidea				
TRIGONALIDAE	75	4	endoparasitoid/ hyperparasitoid	eggs ingested by caterpillar/ sawfly larva but feed only on parasitoid larva in host; also emerge from vespid larvae
Superfamily: Evanioidea				
AULACIDAE	156[3]	20	endoparasitoid	wood-boring beetle and xiphydriid larvae
EVANIIDAE ensign wasps	436[4]	11	predator/ectoparasitoid	eggs in cockroach egg case
GASTERUPTIIDAE	500	15	unknown in North America; in Europe predator/ectoparasitoid	in Europe on solitary bees and wasps
Superfamily: Ceraphronoidea				
CERAPHRONIDAE	360	52	endoparasitoid/ hyperparasitoid	larvae of flies, thrips, lacewings, caterpillars; hyperparasitoid in braconid wasps
MEGASPILIDAE	450	52	parasitoid/ hyperparasitoid	larvae of lacewings, mealybugs, flies, scorpion flies; hyperparasitoid in braconids

FAMILIES IN NORTH AMERICA NORTH OF MEXICO	NUMBER OF WORLD SPECIES	NUMBER OF SPECIES NORTH OF MEXICO	LARVAL FEEDING TYPE	HOSTS
Superfamily: Proctotrupoidea				
HELORIDAE	7	2	endoparasitoid	lacewing larvae
PELECINIDAE	1	1	parasitoid	scarab beetle larvae
PROCTOTRUPIDAE	310	75	endoparasitoid	beetle and fly larvae in soil litter and rotten wood
ROPRONIIDAE	18	3	endoparasitoid	sawfly pupae
VANHORNIIDAE	5	2	parasitoid	wood-boring beetle larvae
Superfamily: Diaprioidea				
DIAPRIIDAE	2300	369		
Belytinae	?	138	endoparasitoid	fly larvae (fungus flies)
Diapriinae	?	222	endoparasitoid	mostly fly larvae and pupae; some beetle larvae; possibly ant larvae
Ismarinae	30	9	hyperparasitoid	dryinid larva in leafhoppers
Superfamily: Platygastroidea				
PLATYGASTRIDAE	4100	530		
Platygastrinae	1100	255	endoparasitoid	cecidomyiid, egg/pupal, egg/larval
Scelioninae	3000	275	endoparasitoid	insect and spider eggs, mostly grasshoppers, crickets; some true bugs; some lepidopterans
Superfamily: Cynipoidea[5]				
CYNIPIDAE gall wasps	2425	640	plant feeder/internal	cause galls on rose, oak; rarely composites; some species inquilines of gall formers
FIGITIDAE	2164	433		
Anacharitinae	125	34	endoparasitoid	green and brown lacewing larvae
Aspiceratinae	204	24	endoparasitoid	syrphid fly pupae
Charipinae	291	55	hyperparasitoid	braconids in aphids, encyrtids in psyllids
Euceroptrinae	4	4	parasitoid	gall wasp in oak galls
Eucoilinae	1322	282	first endo-, then ectoparasitoid	fly larvae/pupae

FAMILIES IN NORTH AMERICA NORTH OF MEXICO	NUMBER OF WORLD SPECIES	NUMBER OF SPECIES NORTH OF MEXICO	LARVAL FEEDING TYPE	HOSTS
Figitinae	205	33	first endo-, then ectoparasitoid	fly larvae/pupae
Thrasorinae	13	1	parasitoid	hymenoptera in *Mimosa* gall
IBALIIDAE	43	11	first endo-, then ectoparasitoid	horntail and anaxelid sawfly larvae in wood
LIOPTERIDAE	161	12		
Lioptrinae	46	3	unknown	unknown
Maryellinae	115	9	unknown	unknown
Superfamily: Chalcidoidea[6]				
AGAONIDAE	774	22		
Agaoninae true fig wasps	368	9	plant feeder	flowers in fig receptacle
Five parasitoid subfamilies	406	13	parasitoids;	it is suspected these species belong to several other chalcidoid families
APHELINIDAE	1311	212	endo-, ecto-, or hyperparasitoids	commonly on scales, mealybugs, whiteflies; rarely lepidopteran eggs
CHALCIDIDAE	1478	140	parasitoid or hyperparasitoids	many different insects
ENCYRTIDAE	4133	490	endo-, ecto-, or hyperparasitoids	many different insects; rarely ticks
EUCHARITIDAE	434	34	ectoparasitoid, planidiform larvae	ant pupae
EULOPHIDAE	5197	824	endo-, ecto-, or hyperparasitoids; introduced plant feeder	many different insects, especially leaf miners; rarely mites; introduced species form galls
EUPELMIDAE	935	119	endo-, ecto-, or hyperparasitoids	many different insects; spider eggs
EURYTOMIDAE	1526	283	ectoparasitoids; some plant feeders	many different insects; spider eggs; some seed feeders, stem gall formers
LEUCOSPIDAE	135	6	ectoparasitoids	larvae of solitary bees and wasps
MYMARIDAE fairyflies	1485	198	endoparasitoids	eggs, including leafhoppers, planthoppers; rarely beetles; some aquatic insects

FAMILIES IN NORTH AMERICA NORTH OF MEXICO	NUMBER OF WORLD SPECIES	NUMBER OF SPECIES NORTH OF MEXICO	LARVAL FEEDING TYPE	HOSTS
ORMYRIDAE	137	17	ecto- and hyperparasitoids	mostly gall-forming wasps and flies; some attack fig wasps
PERILAMPIDAE	292	36	ecto- and hyperparasitoid, planidiform larvae	caterpillars, beetle larvae, grasshoppers, lacewings
PTEROMALIDAE	3627	554	endo-, ecto-, or hyperparasitoids; rarely plant feeder	many different insects; a few gall formers on plants
SIGNIPHORIDAE	78	22	endo- and hyperparasitoid	mostly scales, mealybugs
TANAOSTIGMATIDAE	97	15	plant feeder	galls on stems, leaves, and seeds of various plants
TETRACAMPIDAE	58	10	endo- and ectoparasitoid	fly larvae, eggs of beetles and sawflies
TORYMIDAE	1051	220	ectoparasitoids, plant feeders	gall-forming flies and wasps; solitary bees; internally in seeds
TRICHOGRAMMATIDAE	935	135	endoparasitoids	eggs, including lepidopterans; some aquatic insects
Superfamily: Mymarommatoidea				
MYMAROMMATIDAE	11	2	parasitoid?	possibly barklice eggs
Superfamily: Ichneumonoidea[7]				
BRACONIDAE	17600	2038	endo-, ecto-, rarely hyperparasitoids; rarely gall former (in tropics)	huge number of different insects, mostly caterpillars, beetles, flies
ICHNEUMONIDAE	23000	4190	endo-, ecto-, often hyperparasitoids; rarely cleptoparasitoids in bees (tropics)	huge number of different insects

PREDATORY WASPS, BEES, AND ANTS
(Suborder APOCRITA: ACULEATA or VESPOMORPHA)

Superfamily: Chrysidoidea				
BETHYLIDAE[8]	2393	199		
Bethylinae	457	40	ectoparasitoid	concealed moth larvae, or moves larva to concealed place
Epyrinae	935	98	ectoparasitoid	concealed beetle larvae (for example, among stored grains)

FAMILIES IN NORTH AMERICA NORTH OF MEXICO	NUMBER OF WORLD SPECIES	NUMBER OF SPECIES NORTH OF MEXICO	LARVAL FEEDING TYPE	HOSTS
Pristocerinae	799	61	ectoparasitoid	wood-boring and soil-dwelling beetle larvae; beetles in ant nests
CHRYSIDIDAE[9]	3000	231		
Cleptinae	84	7	ectoparasitoid to host but inside cocoon	on prepupae of sawflies
Amiseginae	108	4	endoparasitoid	eggs of walking sticks
Chrysidinae cuckoo wasps	2200	220	commonly ectoparasitoid, rarely cleptoparasitoid	mature larvae/prepupae of solitary bees and wasps
DRYINIDAE	1100	108		
Anteoninae	250	14	parasitoid sac	leafhoppers
Aphelopinae	55	6	parasitoid sac, rarely endoparasitoid	immature and adult treehoppers and leafhoppers
Bocchinae	60	10	parasitoid sac	leafhoppers and issid planthoppers
Dryininae	190	16	parasitoid sac	fulgorid planthoppers
Gonatopodinae	390	60	parasitoid sac	leafhoppers and fulgorid planthoppers
Plesiodryininae	1	1	unknown	unknown
Thaumatodryininae	22	1	parasitoid sac	flatid planthoppers
EMBOLEMIDAE	10	2	parasitoid sac	achilid planthoppers in fungi
SCLEROGIBBIDAE	10	2	ectoparasitoid	on webspinners (Embioptera)

Superfamily: Apoidea

Digger wasps, hunting wasps, sphecid wasps[10]
Families sometimes placed as Spheciformes within the superfamily

AMPULICIDAE	198	4	predator	cockroach adults
CRABRONIDAE	8661	1335		
Astatinae	153	36	predator	true bugs
Bembicinae	1710	517	predator (rarely cleptoparasitoid on other wasps)	many different hosts, especially true bugs, leafhoppers, planthoppers, flies, and grasshoppers; cicadas, rarely dead insects

FAMILIES IN NORTH AMERICA NORTH OF MEXICO	NUMBER OF WORLD SPECIES	NUMBER OF SPECIES NORTH OF MEXICO	LARVAL FEEDING TYPE	HOSTS
Crabroninae	4616	450	predator	many different insect orders, some spiders
Mellininae	16	7	predator	flies
Pemphredoninae	1025	180	predator	thrips, collembola, leafhoppers, planthoppers, aphids
Philanthinae	1141	7	predator	bees, ants, beetles
SPHECIDAE	711	123		
Ammophilinae	307	81	predator, some cleptoparasitoids	moth and sawfly larvae
Sceliphrinae	144	8	predator	grasshoppers, spiders
Sphecinae	260	34	predator	grasshoppers, cicadas

Bees[11]

Families sometimes placed as Apiformes or Apomorpha
within the superfamily; subfamilies given only for Apidae

FAMILIES IN NORTH AMERICA NORTH OF MEXICO	NUMBER OF WORLD SPECIES	NUMBER OF SPECIES NORTH OF MEXICO	LARVAL FEEDING TYPE	HOSTS
APIDAE	5636	1622		
Apinae digger bees, honey bees, bumble bees (south of U.S./Mexico border: stingless bees, orchid bees)	3441	931	pollen, digger bees solitary; honey bees and stingless bees eusocial; bumble bees mostly social, some social parasitoids; orchid bees mostly solitary; rarely cleptoparasitoids	larval cells in soil (digger bees), others constructed nests (colonies) either underground or in cavities; orchid bees with resin-lined nests in wood, soil, plant stems, or protected overhangs
Nomadinae cuckoo bees	1207	581	pollen, solitary; all clepto-parasitoids of other bees	larval cells in soil
Xylocopinae carpenter bees	988	110	pollen, mostly solitary, some gregarious, some subsocial; rarely social parasitoids	larval cells in wood or plant stems
COLLETIDAE yellow-faced bees; plasterer bees	2498	320	pollen, solitary; rarely cleptoparasitoids	larval cells in soil or in hollow stems
HALICTIDAE sweat bees	4084	828	pollen, solitary, rarely communal or primitively eusocial; some cleptoparasitoids; rarely social parasitoids	larval cells in soil or rotting wood

FAMILIES IN NORTH AMERICA NORTH OF MEXICO	NUMBER OF WORLD SPECIES	NUMBER OF SPECIES NORTH OF MEXICO	LARVAL FEEDING TYPE	HOSTS
MEGACHILIDAE leafcutter bees, mason bees, resin bees, carder bees	3952	861	pollen, solitary, rarely cleptoparasitoids	larval cells in soil, pithy stems, preformed cavities (all lined with plant parts) or formed on rocks, stems, or leaves (from resin, mud, and pebbles)
MELITTIDAE oil-collecting bees	183	32	pollen, solitary	larval cells in soil
Superfamily: Vespoidea				
BRADYNOBAENIDAE	155	48		
Chyphotinae	55	45	unknown	
Typhoctinae	10	3	ectoparasitoid	immature solpugids
MUTILLIDAE velvet ants	5000	420		
Mutillinae	1800	60	ectoparasitoid	mainly larvae of bees and wasps in pupal cases
Myrmosinae	50	15	ectoparasitoid	bee larvae
Sphaeropthalminae	2500	360	ectoparasitoid	larvae of many hosts, usually in pupal cases, including bees, wasps, flies, lepidopterans, beetles, and cockroach egg cases
POMPILIDAE spider wasps	4200	282		
Ceropalinae	200	25	cleptoparasitoids of other pompilids, ectoparasitoids on living host	spiders
Pepsinae	2000	115	predator	spiders
Pompilinae	2000	142	predator, some ectoparasitoids on living host	spiders
RHOPALOSOMATIDAE	34	3	ectoparasitoid	immature crickets
SAPYGIDAE	80	17		
Fedtschenkiinae	?	1	ectoparasitoid	soil-nesting eumenid wasp larvae
Sapyginae	?	16	ectoparasitoid or cleptoparasitoid	cleptoparasitoids of megachilid and anthophorid bee larvae or eumenid wasp larvae
SCOLIIDAE	300	21	ectoparasitoid	scarab or weevil larvae

FAMILIES IN NORTH AMERICA NORTH OF MEXICO	NUMBER OF WORLD SPECIES	NUMBER OF SPECIES NORTH OF MEXICO	LARVAL FEEDING TYPE	HOSTS
SIEROLOMORPHIDAE	10	6	unknown	unknown
TIPHIIDAE[12]	2000	220		
Anthoboscinae	63	1	ectoparasitoid	possibly scarab beetle larvae
Myzininae	340	14	ectoparasitoid	scarab beetle larvae, some on tiger beetle larvae or cerambycid larvae
Methochinae	77	4	ectoparasitoid	tiger beetle larvae
Tiphiinae	623	142	ectoparasitoid	scarab beetle larvae
Brachycistidinae	77	61	ectoparasitoid	scarab beetle larvae
VESPIDAE	4060	315		
Eumeninae	3000	250	predator	mostly caterpillars, less commonly beetle larvae, rarely sawfly larvae
Euparagiinae	10	10	predator	weevil larvae
Masarinae pollen wasps	220	15	pollen	pollen balls
Polistinae paper wasps	700	22	predator, eusocial, some cleptoparasitoids	masticated insects of many types, mainly caterpillars
Vespinae yellow jackets, hornets	80	16	predator, eusocial, some cleptoparasitoids	masticated insects of many types, mainly caterpillars
Ants[13]				
FORMICIDAE ca. 20 world subfamilies	12,513	750		
Amblyoponinae Dracula ants	?	5	predators; eusocial in soil, ground litter	feed on centipedes, caterpillars, soil invertebrates; queens feed on fluids of own larvae
Cerapachyinae	208	5	predators; eusocial in soil	feed on larvae of other ants and possibly termites
Dolichoderinae	1065	40	predators, scavengers, plant feeders; eusocial in soil, litter, hollow plant stems, tree trunks; mound builders; some social parasites on related ants	plant fluids, aphid and scale secretions

FAMILIES IN NORTH AMERICA NORTH OF MEXICO	NUMBER OF WORLD SPECIES	NUMBER OF SPECIES NORTH OF MEXICO	LARVAL FEEDING TYPE	HOSTS
Ecitoninae army ants	193	27	predators; eusocial as nomads, temporarily nest in cavities and subterranean	feed on other arthropods
Ectatomminae	?	2	predator; eusocial in soil	brood of other ants; possibly millipedes
Formicinae carpenter ants, citronella ants, honey ants, mound building ants	?	125	mostly nectar, plant exudates; eusocial in soil, dead wood, trees; many social parasites;	plant exudates, or via plant feeders (including roots) such as aphids, scales; wood fiber
Myrmicinae harvester ants, fire ants, leaf-cutting ants, pharaoh ants	6788	500	seeds, plant parts converted to fungi; predators; some social parasites; eusocial in soil, wood, cartons	masticated plant parts, fungi grown on chewed leaves; insects, including brood of other ants
Ponerinae trap-jaw ants	1701	24	predator, eusocial in soil/rotten logs	masticated insects; some specialize on pillbugs
Proceratiinae	?	9	egg predators; eusocial in soil, leaf litter, rotten logs	eggs of arthropods
Pseudomyrmecinae	226	10	predator, eusocial in hollow plant stems, twigs, logs, soil; some social parasites	masticated insects

[1] World species figures are from ECatSym: Electronic World Catalog of Symphyta Online Version 2.0 (unless noted otherwise; www.zalf.de/home_zalf/institute/dei/php_e/ecatsym/index.html)
[2] Schmidt and Smith (2008)
[3] Smith (2001)
[4] Deans (2005)
[5] Figures, personal communication from Matt Buffington, Systematic Entomology Laboratory, U.S. Department of Agriculture, Washington, D.C.
[6] Figures, personal communication from John Noyes, The Natural History Museum, London
[7] World figures, personal communication from Dicky Yu, University of Kentucky, Lexington; American figures calculated from Nomina Insecta Nearctica (www.nearctica.com/nomina/wasps/hymenop.htm)
[8] World figures, personal communication from Celso Azevedo, Universidade Federal do Espirito Santo, Maruípe, Brazil
[9] Subfamily figures, Kimsey and Bohart (1991)
[10] Figures from Catalog of Sphecidae (http://research.calacademy.org/research/entomology/Entomology_Resources/Hymenoptera/sphecidae/Genera_and_species_PDF/introduction.htm)
[11] Figures from Discover Life, Apoidea Species (www.discoverlife.org/mp/20q?guide=Apoidea_species&flags=HAS:)
[12] World figures, personal communication from Lynn Kimsey, University of California, Davis
[13] Classification and American numbers based on Fisher and Cover (2007); world figures from antbase.org

World Hymenopteran Families Not Found North of Mexico

FAMILY	NUMBER OF WORLD SPECIES	REGION	LARVAL FEEDING TYPE	HOSTS
SAWFLIES (Suborder SYMPHYTA)[1]				
Superfamily: Pamphilioidea				
MEGALODONTESIDAE webspinning sawflies	37	Eurasia	plant feeding/ external	leaves of many herbaceous plants
Superfamily: Tenthredinoidea				
BLASTICOTOMIDAE	13	Eurasia	plant feeding, borers	stems of ferns
PARASITOID WASPS (Suborder APOCRITA: PARASITICA)				
Superfamily: Megalyroidea				
MEGALYRIDAE	45	Australia, South America, Africa, Southeast Asia	probably endoparasitoid	beetle larvae under tree bark; one species attacks wasp larvae
Superfamily: Proctotrupoidea				
AUSTRONIIDAE	3	Australia	unknown	
PERADENIIDAE	2	Australia	unknown	
PROCTORENYXIDAE	1	Australia	unknown	
Superfamily: Diaprioidea				
MONOMACHIDAE	11	Australia, South America	parasitoid	one species known from flies
MAAMINGIDAE	2	New Zealand	unknown	
Superfamily: Cynipoidea				
AUSTROCYNIPIDAE	1	Australia	plant feeder/internal	seeds of *Araucaria*
Superfamily: Chalcidoidea				
ROTOITIDAE	2	New Zealand, South America	unknown	

FAMILY	NUMBER OF WORLD SPECIES	REGION	LARVAL FEEDING TYPE	HOSTS
PREDATORY WASPS, BEES, AND ANTS (Suborder APOCRITA: ACULEATA)				
Superfamily: Chrysidoidea				
PLUMARIIDAE	20	South America, southern Africa	unknown	
SCOLEBYTHIDAE	3	South America, southern Africa, Australia	unknown, but possibly ectoparasitoid	possibly beetles
Superfamily: Apoidea				
Hunting wasps				
HETEROGYNAIDAE	8	Africa, southern Europe	unknown	
Bees				
STENOTRITIDAE	21	Australia	pollen, solitary	larval cells in soil

[1] World species figures are from ECatSym: Electronic World Catalog of Symphyta Online Version 2.0 (www.zalf.de/home_zalf/institute/dei/php_e/ecatsym/index.html)

Some Useful Websites

THOUSANDS OF WEBSITES provide information on various aspects of bees, wasps, and ants. In addition to being numerous, websites can be both highly unstable (that is, disappearing, greatly out of date, or changing addresses) as well as great sources of misinformation. Below I've listed some of the more stable, reliable primary sources for Hymenoptera, which in turn will lead to further adventures into the huge amount of literature on the subject. Unfortunately for gardeners, many of these tend to be technical sites devoted to the dissemination of information for dedicated amateurs as well as scientists. The only society devoted to the study of all Hymenoptera, not just certain groups, is the International Society of Hymenopterists, which is listed below. Some other websites might seem to be important but are not given because the site has been discontinued by its creator (for example, Online Catalog of the North American Ants) or is incomplete or mostly inactive, as indicated by its most current update.

Alternative Pollinators: Native Bees
www.attra.org/attra-pub/nativebee.html (online version)
attra.ncat.org/publication.html (PDF version)
   This is a listing provided by ATTRA, the National Sustainable Agriculture Information Service, from their Master Publication List. Although there are numerous files on just about every aspect of agriculture and home gardening, the site contains only two files on pollinators, the one given above and one on honey bees.

American Beekeeping Federation
www.abfnet.org/
   The American Beekeeping Federation acts on issues affecting the interests and the economic viability of the various sectors of the beekeeping industry. The site provides much ancillary information,

including links, for anyone interested in honey bee production, whether commercial or hobbyist.

antbase.org
www.antbase.org/index.htm

Antbase is a collaborative effort between scientists from around the world, aiming at providing the best possible access to the wealth of information on ants. Antbase now provides for the first time access to all the ant species of the world. As of December 2007, more than 12,000 species were recorded at the site. This is a technical site and a bit clunky to use but is among the foremost ant sites on the internet. Many links to other ant websites are given.

Ant Course
www.calacademy.org/research/entomology/Ant_Course

The Ant Course is taught annually, usually at the Southwestern Research Station in Portal, Arizona. From time to time it changes locations. Ant Course is open to all interested individuals, but priority is given to those students for whom the course will have a significant impact on their research. An entomological background is not required. The course covers ant systematics, ecology, behavioral biology, and conservation.

AntWeb.org
www.antweb.org/

AntWeb provides tools for exploring the diversity and identification of ants. These tools have been developed to encourage the study of ants, facilitate the use of ants in inventory and monitoring programs, and provide ant taxonomists with access to images of type specimens. Currently AntWeb contains information on the ant faunas of several areas in the Nearctic and Malagasy biogeographic regions and global coverage of all ant genera. There are links to world subfamilies, genera, and even species of ants and each includes many photographs.

Aussie Bee
http://www.aussiebee.com.au/index.html

This website is produced by the Australian Native Bee Research Centre, and it serves as a source of information on native bees, including stingless bees. There are numerous printed guides to bees, most of which must be purchased. Some images and videos, as well as identification aides, are available at no charge. There are links to many sites treating Australian native bees.

Bee Course
research.amnh.org/invertzoo/beecourse/

The Bee Course is offered annually at the Southwestern Research Station in Portal, Arizona. The course is designed primarily for botanists, conservation biologists, pollination ecologists, and other biologists whose research, training, or teaching responsibilities require a greater understanding of bee taxonomy. It emphasizes the classification and identification of more than 60 bee genera of North and Central America (both temperate and tropical), and the general information provided is applicable to the global bee fauna. Field trips acquaint participants with collecting and sampling techniques, and associated lab work provides instruction on specimen identification, preparation, and labeling.

Bees, Wasps and Ants Recording Society
www.bwars.com

The Bees, Wasps and Ants Recording Society (BWARS) has over 360 members, mainly in Britain and Ireland, but there are also corresponding members from around the world. The society's main focus is to record where the various species of bees, wasps, and ants occur in the United Kingdom. As a result of this recording work, BWARS publishes provisional atlases (free to members) of selected species. Members receive a biannual newsletter containing color photographs, information on meetings, pilot distribution maps, and profiles of bees, wasps, and ants.

Brackenridge Field Laboratory at The University of Texas at Austin
http://web.biosci.utexas.edu/fireant/FAQ.html

Many questions concerning fire ants are addressed at this site, ranging from how to identify them, what effects they have on the environment, and what to do about them.

Catalog of Sphecidae sensu lato

http://research.calacademy.org/research/entomology/Entomology_
Resources/Hymenoptera/sphecidae/Genera_and_species_PDF/
introduction.htm

> This site consists of a constantly updated catalog of information
> about three families of predatory wasps, Ampulicidae, Sphecidae,
> and Crabronidae, all of which used to be classified as Sphecidae. It is
> maintained single-handedly by Dr. Wojciech J. Pulawski (California
> Academy of Sciences). Of special note is the selection called "Number
> of Species" (http://research.calacademy.org/research/entomology/
> Entomology_Resources/Hymenoptera/sphecidae/Number_of_
> Species.htm), which gives an accurate tally of world species of this
> group.

Chrysis.net

www.chrysis.net/index_en.php

> This site treats the family Chrysididae (cuckoo wasps), which con-
> tains about 3000 described world species in 86 genera. Although the
> emphasis is on Italian species, there is a great amount of general infor-
> mation pertaining to biology, some photographs, and a bibliographic
> database of the entire world literature on the family.

Discover Life

www.discoverlife.org/

> This site is under constant development and provides interactive, pic-
> torial identification guides (www.discoverlife.org/mp/20q) for many
> different sorts of organisms including insects, plants, fungi, and verte-
> brates. For the computer geek, these are actually somewhat fun to use.
> Because guides are in various states of development, they differ widely
> in scope and detail and are not available for every organism even
> within a group. For example, within Hymenoptera there are pictorial
> guides to species of North American bumble bees (www.discoverlife.
> org/mp/20q?guide=Bumblebees) and some ants (www.discoverlife.
> org/mp/20q?guide=Ants). There are guides to bee genera of eastern
> North America, which require that you then go to the genus guide to
> determine species. The most useful general area of the site is an inter-
> active listing to world "Apoidea species" (that is, bees) that provides a

great amount of information if you have some notion of what you are looking for (www.discoverlife.org/mp/20q?guide=Apoidea_species& flags=HAS:). This is an extremely valuable tool, but might be a bit difficult for the nonspecialist to use.

Earthlife
www.earthlife.net/

This is a somewhat eclectic site about many different kinds of animals. The Hymenoptera section, called "Gordon's Hymenoptera Page" (www.earthlife.net/insects/hymenop.html), is worth a visit.

ECatSym: Electronic World Catalog of Symphyta Online Version 2.0
www.zalf.de/home_zalf/institute/dei/php_e/ecatsym/index.html

In 2008 this consisted of a searchable database of the world's 801 valid sawfly genera and 8090 valid species within 28 families including fossils. It is the source for all things symphytan.

Ecoregional Planting Guides
www.pollinator.org/guides.htm

This is a special section of the Pollinator Partnership site (see below). This site is especially useful for gardeners who wish to design gardens to attract specific pollinators based on where they live. It features downloadable PDF planting guides to 35 U.S. ecoregions, selectable by zip code. Each guide provides plant-pollinator information specific to that ecoregion, including bloom periods, native plants that attract pollinators, and habitat hints, as well as additional resources and tips, including habitat and nesting requirements of different pollinators, a basic checklist, and where to access additional information.

Evanioidea Online
http://evanioidea.info/public/site/evanioidea/home

This site is a catalog of information about evanioid wasps. The site treats three parasitic families in detail: Evaniidae, Aulacidae, and Gasteruptiidae. Information about each family is reached through links on the home page. Although mostly technical, there are some useful bits of biological information via the links.

Fauna Europaea
http://www.faunaeur.org/

This is a searchable database of all scientific names (select Data Services) of multicellular European animals. Categories such as family, genus, or species can be entered and lists of corresponding taxa are presented along with tables and distribution maps.

Genera Ichneumonorum Nearcticae
http://www.amentinst.org/GIN/

Genera Ichneumonorum Nearcticae aims to present concise current information about the taxonomy, biology, distribution, and species richness of North American Ichneumonidae. The sections listed on the side-bar provide an introduction to each subfamily present in the Nearctic region, together with a concise subfamilial diagnosis and a synopsis of its biology. Each section is completed by keys to Nearctic genera, with each key augmented by high-quality photographs, a summary of species present, and reference to the current keys to species.

Handbook of Nearctic Chalcidoidea
www.codex.begoniasociety.org/chalcidkey/

This is an interactive, pictorial key to all known chalcidoid families (except the obscure Rotoitidae from New Zealand and Chile). In addition, much information about each family is summarized, including biology.

Help the Honeybees
www.helpthehoneybees.com

This is actually a fun and entertaining animated site appropriate for youngsters, but it might require a bit of reading help from adults. Kids can make a bee fly around a field examining various kinds of plants and their pollination requirements. They can even make a custom bee and email it to their friends. The site is sponsored by Häagen-Dazs ice cream.

HymAToL (Hymenoptera—Assembling the Tree of Life)
www.hymatol.org

HymAToL is a project funded by the National Science Foundation under the "Assembling the Tree of Life" initiative, which aims to reconstruct the evolutionary history of all organisms. The goal is to construct a large-scale phylogeny of the Hymenoptera. Approximately 115,000 species have already been described and estimates for the total number of species range up to 2.5 million. Although a highly technical site, there is a list of families with photographs that pop up by use of the cursor (use "Introduction" at left of screen), and there are some links to internet identification keys that may be of interest to the more technically inclined. Currently the keys are accessible only to PC users.

Hym Course
http://hymcourse.org/default_new.asp?action=Show_Home
This is a one-week workshop providing participants with knowledge and experience in identifying parasitic Hymenoptera, stinging wasps (aculeates), sawflies, and limited information on ants and bees. Aspects of biology, collecting, and preserving Hymenoptera are also covered.

Hymenoptera (wasps, ants, bees): Iowa State Entomology Index of Internet Resources
www.ent.iastate.edu/list/directory/89/vid/5
This is a list and link resource to a large number of websites devoted to wasps, ants, and bees.

Hymenoptera Interactive Keys
sharkeylab.org/sharkeylab/sharkeyKeys.php
This is an illustrated, interactive key to families of Hymenoptera. It requires a software download and works only on PCs. In addition to families, nine superfamilies are also featured, and keys to subfamilies and genera of New World Braconidae are included. The latter keys are presented in English and Spanish.

Hymenoptera Online Database
http://osuc.biosci.ohio-state.edu/HymOnline/
The site is useful for checking species names and for illustrations of some families of Hymenoptera, but it is a work in progress most useful to a specialist.

International Bee Research Association
www.ibra.org.uk/
> This site is dedicated to the honey bee and provides as much information as one could use as well as links to numerous other sites for honey bees. It is oriented toward Europe. The U.S. equivalent is the American Beekeeping Federation (see above).

International Society of Hymenopterists
www.hymenopterists.org
> This is the only world body representing the entire range of Hymenoptera information. Many of the site's features may be used by non-members, and there are especially good links to other sites that treat Hymenoptera. The society publishes the twice-yearly *Journal of Hymenoptera Research*, a highly technical journal.

The Lurker's Guide to Leafcutter Ants
www.blueboard.com/leafcutters/
> A. San Juan's guide to everything you'd want to know about the lives of leafcutter ants. The site includes many images and links.

Mutillidae: Velvet Ants
http://www.biology.usu.edu/labsites/pompilidweb/Mutillidae.htm
> This is a companion website to the one on spider wasps (Pompilidae) listed below.

Nomina Insecta Nearctica
www.nearctica.com/nomina/wasps/hymenop.htm
> This website provides a searchable list of all described species of Hymenoptera from North America north of Mexico. There are more than 17,000 species names, and the main difficulty is that one must first know the family to which a species belongs before it can be searched for. The site is of some use if you want to know how many species of a family or genus there are. Unfortunately, it is valid only to 1979, so none of the hundreds of species that have been described since then are included.

North American Pollinator Protection Campaign
www.nappc.org

The NAPPC consists of more than 90 affiliated organizations working to implement, promote, and support a clear, continent-wide coordinated action plan to increase the numbers and health of both resident and migratory pollinating animals in North America. Most of those organizations can be accessed via links on this website.

Pergidae of the World
http://www.pergidae.net

The site contains an online catalog of the sawfly family Pergidae (Symphyta), covering some 434 world species as of December 2008. In addition to taxonomic information, there is information on biology, food plants, and distribution.

Plants Attractive to Native Bees
http://www.ars.usda.gov/Main/docs.htm?docid=12052

This site is part of the Pollinating Insect Research Lab, U.S. Department of Agriculture, Agricultural Research Service, Logan, Utah (www.ars.usda.gov/main/docs.htm?docid=5609), a lab that specializes in the role of bees in the environment. In addition to listing bee-attracting plants, there are a number of selections under "Products and Services" concerning bees, including "Gardening for Native Bees in North America," "Building a Nesting Block," and "ID a Bumblebee."

Pollinator-Friendly Cut Flower Plants
www.smallfarmsuccess.info/poll_friendly.cfm

The site offers a downloadable list of pollinator-friendly perennials and annuals, with descriptions of plant height, flower color, seasonality, cultural notes, and sources.

Pollinator Partnership
www.pollinator.org

Nearly 80 percent of our world's crop plants require pollination, a function that is vital for plant reproduction. Whether you are a gardener, a farmer, a resource manager, an educator, or simply an interested consumer, the Pollinator Partnership provides news, resources, programs, and an extensive digital library to support you in helping pollinators.

Pompilidae: Spider Wasps
http://www.biology.usu.edu/labsites/pompilidweb/
This site is devoted to information about a single family of wasps, containing some 5000 species, that collect spiders as prey for their young.

Universal Chalcidoidea Database
http://www.nhm.ac.uk/research-curation/research/projects/chalcidoids/
This is a searchable database for some 25,000 species of wasps in the superfamily Chalcidoidea. It is designed more for the specialist than the layman. Information includes more than 120,000 host/associate records (including associations with food plants of the hosts) and more than 140,000 distribution records. A bibliographic database lists more than 40,000 references regarding Chalcidoidea. More than 350 images of a wide range of living chalcidoids are also available. The website is the brainchild of John Noyes, The Natural History Museum, London.

Urban Bee Gardens
nature.berkeley.edu/urbanbeegardens/general_solitarybees.html
Centered at the University of California, Berkeley, this site has information of a general nature for those interested in various aspects of bees and their role in the garden. There are planting guides to attract bees, as well as discussions and images of bees and wasps.

World Bee Checklist
www.itis.gov/beechecklist.html
This is a subset of listings from the Integrated Taxonomic Information System. There are several useful databases here, but the simplest to download (in Excel format) is also the least conspicuous, being the last choice of four. It simply lists the world's valid bee species alphabetically, first by family, then genus and species and the subfamilies to which they belong. There is no biological or distributional information. Its practical use is limited to searching for a genus or species name either to determine correct spelling or to find the family name

if one only knows the genus and species. The Discover Life (listed above) coverage of "Apoidea species" (bee species) is much more useful, in my opinion, than this one.

World Braconidae Name Index
http://www.sharkeylab.org/bracnames/index.php

This is a list of names in the family Braconidae extracted from the next list (that is, World Ichneumonoidea) and updated to 2007. It provides an alphabetized list of nearly 24,000 scientific names published to the present. Some of these names are no longer considered valid; you can look up a name and if it is no longer the correct scientific name, you will be referred to the valid name.

World Ichneumonoidea
www.taxapad.com/

This site provides a searchable list of names of more than 50,000 species of the superfamily Ichneumonoidea. The data are updated through 2004.

Xerces Society
http://www.xerces.org/

The Xerces Society is an international nonprofit organization dedicated to protecting biological diversity through invertebrate conservation. The site has a section dedicated to pollinator conservation (http://www.xerces.org/pollinator-conservation/) and to several useful publications about pollinators (http://www.xerces.org/fact-sheets/). Of particular interest are the downloadable PDFs "Nests for Native Bees," "Plants for Native Bees in the Pacific Northwest," "Plants for Native Bees in the Upper Midwest," "Plants for Native Bees in California," and "Plants for Native Bees in North America."

References

INCLUDED IN THIS list are all the references cited specifically within the text, as well as a few general references that provide additional reading. In some cases the references are quite enjoyable, but in others they are scientific papers, best read by those who need to know the origin of information because they either do not believe what I've written, or they enjoy slogging through dense, technical material. Some websites are cited here, rather than in the list of websites, because they have information of a specific nature that was cited in the text.

Alvarez, J. M. 1997. Chapter 26: Largest parasitoid brood. *University of Florida Book of Insect Records*, http://entnemdept.ufl.edu/walker/ufbir/chapters/chapter_26.shtml.

Anderson, C. 1999. Ant palaces. *Outside In: The Missouri Conservationist for Kids Online*, http://www.mdc.mo.gov/kids/out-in/1999/2/2.html.

Arnett, R. H., and R. L. Jacques Jr. 1981. *Simon & Schuster's Guide to Insects*, 2nd ed. New York, Fireside Books.

Askew, R. R. 1971. *Parasitic Insects*. New York, American Elsevier.

Austin, A. D., and M. Dowton, eds. 2000. *Hymenoptera: Evolution, Biodiversity and Biological Control*. Canberra, CSIRO Publications.

Banks, B. E. C., and R. A. Shipolini. 1986. Chemistry and pharmacology of honey-bee venom, p. 329–416. In T. Piek, ed. *Venoms of the Hymenoptera*. London, Academic Press.

BBC News. 2004a. Super ant colony hits Australia, news.bbc.co.uk/2/hi/science/nature/3561352.stm.

BBC News. 2004b. Why ants make great gardeners, news.bbc.co.uk/2/hi/science/nature/3499842.stm.

Bellmann, H. 2005. *Bienen, Wespen, Ameisen: Hautflügler Europas*, 2nd ed. [Bees, Wasps, and Ants of Central Europe]. Stuttgart, Kosmos.

Bellmann, H. 2009. *Guide des abeilles, bourdons, guêpes et fourmis d'Europe* [Guide to Bees, Bumble Bees, Wasps, and Ants of Europe]. Lausanne, Delachaux et Neieslte.

Bennett, A. M. R. 2008. Global diversity of hymenopterans (Hymenoptera; Insecta) in freshwater. Hydrobiologia 595: 529–534; PDF available online at www.springerlink.com/content/1642222383141401/.

Berenbaum, M. 1996. *Bugs in the System*. Reading, Mass., Addison-Wesley.

Betts, C., and D. Laffoley, eds. 1986. *The Hymenopterist's Handbook*, 2nd ed. London, The Amateur Entomologists' Society.

Black, R. 2008. Bees get plants' pests in a flap. BBC News, Science and Environment, http://news.bbc.co.uk/go/pr/fr/-/2/hi/science/nature/7796138.stm.

Bohart, R. M., and A. S. Menke. 1976. *Sphecid Wasps of the World*. Berkeley, University of California Press.

Borror, D. J., and R. E. White. 1998. *A Field Guide to Insects* (Peterson Field Guides), 2nd rev. ed. Boston, Houghton Mifflin.

Bosch, J., and W. Kemp. 2001. *How to Manage the Blue Orchard Bee: Handbook Series Book 5*. Beltsville, Md., Sustainable Agriculture Network; out of print, but PDF available online at www.sare.org/publications/bob.htm.

Boucek, Z. 1997. Chapter 4: Agaonidae. *In* G. A. P. Gibson, J. T. Huber, and J. B. Woolley, eds. *Annotated Keys to the Genera of Nearctic Chalcidoidea (Hymenoptera)*. Ottawa, National Research Council of Canada.

Buchmann, S. L., and G. P. Nabhan. 1997. *The Forgotten Pollinators*. Washington, D.C., Island Press.

Cabrera-Mireles, H. 1998. Chapter 36: Most polyandrous. *University of Florida Book of Insect Records*, http://entnemdept.ufl.edu/walker/ufbir/chapters/chapter_36.shtml.

Cane, J. 2005. Squash Pollinators of the Americas Survey (SPAS), www.ars.usda.gov/Research/docs.htm?docid=16595.

Carson, R. 1962. *Silent Spring*. Boston, Houghton Mifflin.

Castner, J. L. 2001. *Photographic Atlas of Entomology and Guide to Insect Identification*. Gainesville, Fla., Feline Press.

Castner, J. L. 2004. *Photographic Atlas of Botany and Guide to Plant Identification*. Gainesville, Fla., Feline Press.

Centers for Disease Control and Prevention. 1997. Dog-bite-related fatalities: United States, 1995–1996. *Morbidity and Mortality Weekly Report* 46(21): 463–466, www.cdc.gov/mmwr/preview/mmwrhtml/00047723.htm.

Chinery, M. 1993. *Collins Field Guide: Insects of Britain and Northern Europe*, 3rd ed. London, Harper Collins.

Cory, E. N., and E. E. Haviland. 1938. Population studies of *Formica exsectoides* Forel. *Annals of the Entomological Society of America* 31: 50–57.

Cranshaw, W. 2004. *Garden Insects of North America: The Ultimate Guide to Backyard Bugs* (Princeton Field Guides). Princeton, N.J., Princeton University Press.

Crompton, J. 1955. *The Hunting Wasp*. Cambridge, Mass., The Riverside Press.

Cushing, P. E., and R. G. Santangelo. 2002. Notes on the natural history and hunting behavior of an ant-eating zodariid spider (Arachnida, Araneae) in Colorado. *The Journal of Arachnology* 30: 618–621.

Danforth, B. N., J. Fang, S. Sipes, S. G. Brady, and E. Almeida. 2004. Phylogeny and molecular systematics of bees (Hymenoptera: Apoidea). Ithaca, N.Y., Cornell University, http://www.entomology.cornell.edu/BeePhylogeny/.

Darwin, C. 1860. Letter to Asa Gray, May 22, 1860. *In* F. Darwin, ed. *The Life and Letters of Charles Darwin*, Vol. 2. New York, Appleton; available at Darwin Correspondence Project: www.darwinproject.ac.uk/darwinletters/calendar/entry-2814.html.

Dawkins, R. 1995. *River out of Eden: A Darwinian View of Life*. New York, Basic Books.

Deans, A. R. 2005. Annotated catalog of the world's ensign wasp species (Hymenoptera: Evaniidae). *Contributions of the American Entomological Institute* 34: 1–165; PDF available online at http://evanioidea.info/public/site/evanioidea/home/about.

Eaton, E. R. 2005. Comments at Bugguide. Bugguide.net/node/view/2#2A82DC.

Eaton, E. R., and K. Kaufman. 2007. *Kaufman Field Guide to Insects of North America*. Boston, Houghton Mifflin.

Edwards, M., and M. Jenner. 2009. *Field Guide to the Bumblebees of Great Britain and Ireland*, 2nd rev. ed. Eastbourne, Ocelli Limited.

Engel, M. S. 1999. The taxonomy of recent and fossil honey bees (Hymenoptera: Apidae; *Apis*). *Journal of Hymenoptera Research* 8: 165–196.

Evans, A. V. 2007. *National Wildlife Federation Field Guide to Insects and Spiders and Related Species of North America*. New York, Sterling.

Evans, H. E. 1964. *Wasp Farm*. London, George G. Harrap & Co., Ltd.

Evans, H. E. 1966. *The Comparative Ethology and Evolution of the Sand Wasps*. Cambridge, Mass., Harvard University Press.

Evans, H. E. 2001. *A Naturalist's Years in the Rocky Mountains*. Boulder, Colo., Johnson Books.

Evans, H. E., and M. J. West Eberhard. 1970. *The Wasps*. Ann Arbor, University of Michigan Press.

Evans, M. A., ed. 2005. *The Man Who Loved Wasps*. Boulder, Colo., Johnson Books.

Fisher, B. L., and S. P. Cover. 2007. *Ants of North America: A Guide to the Genera*. Berkeley, University of California Press.

Freeland, W. J., and W. J. Boulton. 1992. Coevolution of food webs: parasites, predators and plant secondary compounds. *Biotropica* 24: 309–327.

Gahlhoff, J. E. 1998. Chapter 38: Smallest adult. *University of Florida Book of Insect Records*, http://entnemdept.ufl.edu/walker/ufbir/chapters/chapter_38.shtml.

Gauld, I., and B. Bolton, eds. 1988. *The Hymenoptera*. Oxford, Oxford University Press.

Gauld, I., and K. J. Gaston. 1995. The Costa Rican Hymenoptera fauna, p. 13–19. In P. Hanson and I. D. Gauld, eds. *The Hymenoptera of Costa Rica*. Oxford, Oxford University Press.

Gibbons, B. 1996. *Field Guide to Insects of Great Britain and Northern Europe*. Wiltshire, Crowood Press Ltd.

Gibson, G. A. P., J. T. Huber, and J. B. Woolley, eds. 1997. *Annotated Keys to the Genera of Nearctic Chalcidoidea (Hymenoptera)*. Ottawa, National Research Council of Canada.

Giraud, T., J. S. Pedersen, and L. Keller. 2002. Evolution of supercolonies: the Argentine ants of southern Europe. *Proceedings of the National Academy of Sciences of the United States of America* 99: 6075–6079;

PDF available at http://www.pnas.org/content/99/9/6075.full.
pdf+html.

Gorb, E., and S. Gorb. 2003. *Seed Dispersal by Ants in a Deciduous Forest Ecosystem: Mechanisms, Strategies, Adaptations.* New York, Springer.

Goulet, H., and J. T. Huber, eds. 1993. *Hymenoptera of the World: An Identification Guide to Families.* Ottawa, Canada Communications Group Publishing.

Greathead, D. J. 1986. Parasitoids in classical biological control, p. 289–318. In J. Waage and D. Greathead, eds. *Insect Parasitoids.* New York, Academic Press.

Grissell, E. E. 1999. Hymenopteran biodiversity: some alien notions. *American Entomologist* 45: 235–244.

Grissell, E. E. 2001. *Insects and Gardens: In Pursuit of a Garden Ecology.* Portland, Ore., Timber Press.

Grissell, E. E., and M. E. Schauff. 1997. *A Handbook of the Families of Nearctic Chalcidoidea (Hymenoptera),* 2nd rev. ed. Washington, D.C., Entomological Society of Washington, Handbook 1: 1–87; available online at http://www.sel.barc.usda.gov/hym/chalcid.html.

Herzberg, L. 1996. *A Pocketful of Galls: William Brodie and the Natural History Society of Toronto.* Toronto, privately printed by the author.

Hoddle, M. S. 2003. Classical biological control of arthropods in the 21st century, p. 3–16. In R. G. Van Driesche, ed. *Proceedings of the First International Symposium on Biological Control of Arthropods,* Honolulu, Hawaii, January 14–18, 2002. U.S. Department of Agriculture Forest Service, publication FHTET-03-05.

Hölldobler, B., and E. O. Wilson. 1990. *The Ants.* Cambridge, Mass., Belknap Press.

Hölldobler, B., and E. O. Wilson. 1994. *Journey to the Ants: A Story of Scientific Exploration.* Cambridge, Mass., Belknap Press.

Hölldobler, B., and E. O. Wilson. 2009. *The Superorganism: The Beauty, Elegance, and Strangeness of Insect Societies.* New York, W. W. Norton.

Holliday, C. 2008. Prof. Chuck Holliday's cicada killer page, ww2.lafayette.edu/~hollidac/cicadakillerhome.html.

Horn, T. 2006. *Bees in America: How the Honey Bee Shaped a Nation.* Lexington, University of Kentucky Press.

Kearns, C. A., and J. D. Thomson. 2001. *The Natural History of Bumble-bees*. Boulder, University Press of Colorado.

Kimsey, L. S., and R. M. Bohart. 1991. *The Chrysidid Wasps of the World*. Oxford, Oxford University Press.

Knudsen, J. T., R. Eriksson, J. Gershenzon, and B. Ståhl. 2006. Diversity and distribution of floral scent. *The Botanical Review* 72: 1–120.

Krombein, K. V., P. D. Hurd, D. R. Smith, and B. D. Burks, eds. 1979. *Catalog of Hymenoptera of America North of Mexico*. Washington, D.C., Smithsonian Institution Press.

LaSalle, J., and I. D. Gauld. 1991. Parasitic Hymenoptera and the biodiversity crisis. Redia 74(3): 315–334 (appendix).

LaSalle, J., and I. D. Gauld, eds. 1993. *Hymenoptera and Biodiversity*. Wallingford, U.K., CAB International.

Leahy, C. 1998. *Peterson First Guide to Insects of North America*, 2nd rev. ed. Boston, Houghton Mifflin.

Levick, N. R., J. O. Schmidt, J. Harrison, G. S. Smith, and K. D. Winkel. 2000. Review of bee and wasp sting injuries in Australia and the USA, p. 437–447. In A. D. Austin and M. Dowton, eds. *Hymenoptera: Evolution, Biodiversity and Biological Control*. Canberra, CSIRO Publications.

Lochhead, C. 2008. Un-busy bees a disaster for almost everyone. *San Francisco Chronicle*, June 27, 2008, www.sfgate.com/cgi-bin/article. cgi?f=/c/a/2008/06/27/MNLA11FN5B.DTL.

Losey, J. E., and M. Vaughan. 2006. The economic value of ecological services provided by insects. *BioScience* 56: 311–323.

Meyer, W. L. 1996. Chapter 23: Most toxic insect venom. *University of Florida Book of Insect Records*, http://entnemdept.ufl.edu/walker/ufbir/chapters/chapter_23.shtml.

Michener, C. 2007. *The Bees of the World*, 2nd ed. Baltimore, Md., Johns Hopkins University Press.

National Audubon Society. 1980. *National Audubon Society Field Guide to North American Insects and Spiders*. New York, Knopf.

North American Pollinator Protection Campaign. 2006. Bee importation white paper, www.pollinator.org/Resources/BEEIMPORTATION_AUG2006.pdf.

Olkowski, W., S. Daar, and H. Olkowski. 1994. *Common-Sense Pest Control*. Newtown, Conn., The Taunton Press.

O'Neill, K. M. 2001. *Solitary Wasps: Behavior and Natural History.* Cornell Series in Arthropod Biology. Ithaca, N.Y., Comstock.

O'Toole, C. 1998. *Alien Empire.* London, BBC Books.

O'Toole, C., and A. Raw. 1999. *Bees of the World.* London, Blandford.

Owen, J. 1983. *Garden Life.* London, Chatto and Windus.

Patek, S. N., J. E. Baio, B. L. Fisher, and A. V. Suarez. 2006. Multifunctionality and mechanical origins: ballistic jaw propulsion in trap-jaw ants. *Proceedings of the National Academy of Sciences of the United States of America* 103(34): 12787–12792.

Pope, A. 1734. *Essay on Man, Epistle 1, On the Nature and State of Man with Respect to the Universe.* Available online at http://classiclit.about.com/library/bl-etexts/apope/bl-apope-essay-1.htm.

Preston, C. 2006. *Bee.* London, Reaktion Books, Ltd.

Prys-Jones, O. E., and S. A. Corbet. 1991. *Bumblebees.* Naturalists' Handbook Series. Cambridge, U.K., The Company of Biologists.

Redfern, M., and R. R. Askew. 1992. *Plant Galls.* Naturalists' Handbook Series. Cambridge, U.K., The Company of Biologists.

San Juan, A. 2005. The lurker's guide to leaf cutter ants, http://www.blueboard.com/leafcutters.

Schmidt, J. O. 1986. Allergy to Hymenoptera venoms, p. 509–546. In T. Piek, ed. *Venoms of the Hymenoptera.* London, Academic Press.

Schmidt, J. O. 1990. Hymenoptera venoms: striving toward the ultimate defense against vertebrates, p. 387–419. In D. L. Evans and J. O. Schmidt, eds. *Insect Defenses: Adaptive Mechanisms and Strategies of Prey and Predators.* Albany, State University of New York Press.

Schmidt, S., and D. R. Smith. 2008. Pergidae of the world: an online catalogue of the sawfly family Pergidae (Insecta, Hymenoptera, Symphyta), www.pergidae.net.

Schneider, G. H. 1997. Chapter 25: Greatest host specificity. *University of Florida Book of Insect Records*, http://entnemdept.ufl.edu/walker/ufbir/chapters/chapter_25.shtml.

Science Daily. 2008. Economic value of insect pollination worldwide estimated at U.S. $217 billion. 15 September 2008, www.sciencedaily.com/releases/2008/09/080915122725.htm.

Shafer, G. D. 1949. *The Ways of a Mud Dauber.* Stanford, Calif., Stanford University Press.

Sharkey, M. J. 2007. Phylogeny and classification of Hymenoptera, p. 521–548. In Z.-Q. Zhang and W. A. Shear, eds. *Linnaeus Tercente-*

nary: Progress in Invertebrate Taxonomy. Zootaxa 1668; PDF available at www.mapress.com/zootaxa/2007f/zt01668p548.pdf.

Shepherd, M., S. L. Buchmann, M. Vaughan, and S. H. Black. 2003. *Pollinator Conservation Handbook*. Portland, Ore., The Xerces Society.

Sherwood, V. 1996. Chapter 21: Most heat tolerant. *University of Florida Book of Insect Records*, http://entnemdept.ufl.edu/walker/ufbir/chapters/chapter_21.shtml.

Sieglaff, D. 1994. Chapter 8: Most spectacular mating. *University of Florida Book of Insect Records*, http://entnemdept.ufl.edu/walker/ufbir/chapters/chapter_08.shtml.

Skinner, G. J., and G. W. Allen. 1996. *Ants*. Naturalists' Handbook Series. Cambridge, U.K., The Company of Biologists.

Sleigh, C. *Ant*. 2003. London, Reaktion Books, Ltd.

Smith, D. R. 1993. Systematics, life history, and distribution of sawflies, p. 3–32. *In* M. Wagner and K. F. Raffa, eds. *Sawfly Life History Adaptations to Woody Plants*. San Diego, Academic Press.

Smith, D. R. 2001. World catalog of the family Aulacidae (Hymenoptera). *Contributions on Entomology, International* 4: 263–319.

Stork, N. E. 1996. Measuring global biodiversity and its decline, p. 41–68. In M. L. Reaka-Kudla, D. E. Wilson, and E. O. Wilson, eds. *Biodiversity II*. Washington, D.C., Joseph Henry Press.

Tallamy, D. W. 2009. *Bringing Nature Home: How You Can Sustain Wildlife with Native Plants*. Portland, Ore., Timber Press.

Townes, H. 1975. The parasitic Hymenoptera with the longest ovipositors, with descriptions of two new Ichneumonidae. *Entomological News* 86: 123–127.

Triplehorn, C. A., and N. F. Johnson. 2005. *Borror and DeLong's Introduction to the Study of Insects*, 7th ed. Belmont, Calif., Thomson-Brooks/Cole.

Tuell, J. K., J. S. Ascher, and R. Isaacs. 2009. Wild bees (Hymenoptera: Apoidea: Anthophila) of the Michigan highbush blueberry agroecosystem. *Annals of the Entomological Society of America* 102(2): 275–287.

Vicidomini, S. 2005. Chapter 40: Largest eggs. *University of Florida Book of Insect Records*, http://entnemdept.ufl.edu/walker/ufbir/chapters/chapter_40.shtml.

von Frisch, K. 1953. *The Dancing Bee*. (First published in 1927 in German; this is a translation of the 5th rev. ed.) New York, Harcourt, Brace and Co./Harvest.

Waldbauer, G. 1998. *The Handy Bug Answer Book*. Detroit, Visible Ink Press.

Waldbauer, G. 2003. *What Good Are Bugs? Insects in the Web of Life*. Cambridge, Mass., Harvard University Press.

Weiss, K. 2002. *The Little Book of Bees*. New York, Copernicus Books.

Westcott, C. 1973. *The Gardener's Bug Book*, 4th ed. Garden City, N.Y., Doubleday & Co., Inc.

Wharton, R. A., and P. E. Hanson. 2005. Biology and evolution of braconid gall wasps (Hymenoptera). *In* A. Raman, C. W. Schaefer, and T. M. Withers, eds. *Biology, Ecology, and Evolution of Gall-Inducing Arthropods*. Enfield, N.H., Science Publishers.

Wilson, E. O. 1987. The little things that run the world. *Conservation Biology* 1: 344–346.

Yeo, P. F., and S. A. Corbet. 1995. *Solitary Wasps*, 2nd rev. ed. Naturalists' Handbook Series. Cambridge, U.K., The Company of Biologists.

Index